TRANSDUCERS FOR INSTRUMENTATION

TRANSDUCERS

For

INSTRUMENTATION

By

M.G. JOSHI

Formerly, Scientist at Central Scientific
Instruments Organisation, Chandigarh
Panchkula

UNIVERSITY SCIENCE PRESS

(An Imprint of Laxmi Publications Pvt. Ltd.)

An ISO 9001:2015 Company

BENGALURU • CHENNAI • GUWAHATI • HYDERABAD • JALANDHAR
KOCHI • KOLKATA • LUCKNOW • MUMBAI • RANCHI
NEW DELHI

TRANSDUCERS FOR INSTRUMENTATION

Printed and bound in India
First Edition : 2006, Reprint : 2008, 2011, 2014, 2017, 2020, 2022
UTI-9238-195-TRANSDUCERS FOR INSTRU-JOS
ISBN 978-81-318-0473-5
Price : ₹ **195.00**

Branches

Bengaluru	080-26 75 69 30
Chennai	044-24 34 47 26
Guwahati	0361-254 36 69
Hyderabad	040-27 55 53 83
Jalandhar	0181-222 12 72
Kochi	0484-405 13 03
Kolkata	033-40 04 77 79
Lucknow	0522-430 36 13
Ranchi	0651-224 24 64

PUBLISHED IN INDIA BY
UNIVERSITY SCIENCE PRESS
(An Imprint of Laxmi Publications Pvt. Ltd.)
An ISO 9001:2015 Company
113, GOLDEN HOUSE, GURUDWARA ROAD, DARYAGANJ,
NEW DELHI - 110002, INDIA
Telephone : 91-11-4353 2500, 4353 2501
www.laxmipublications.com
info@laxmipublications.com

C—R/023/09
Printed at : Mehra Offset Printers, Delhi

Dedicated to the memory of my late father, Shri Bal Mukand Joshi

ABOUT THE AUTHOR

M.G. Joshi (born 1935), did his M. Tech in applied Physics from the University of Calcutta way back in 1961, with specialisation in Instrumentation and Instrument technology. Starting his professional career as engineer in Maharashtra State Electricity Board and after a brief stint in private sector as Instrument Engineer, he joined CSIR as a scientist in 1969 and was based at Central Scientific Instruments Organisation, Chandigarh till retirement on superannuation in Feb. 1995. It was a long inning in R & D, management, planning and scientific publication activities.

After retirement, the author pursued a teaching assignment to B. Tech (Instrumentation) students at SLIET, Longowal (Punjab) as a visiting faculty on regular basis followed by a part time lecture assignment to M. Tech (Instrumentation) students at T.T.T.I., Chandigarh.

He is a life member (fellow) of the Institution of Electronics and Telecommunication Engineers (India).

PREFACE

Transducer is essentially a device that provides a usable output in response to a specific measurand. It converts or transduces energy from one form to another. The transducers are generally low powered machines devised to convert the physical variables into a form more suitable for measurement and control. These are the first in a sequence or chain of instrumentation that follows and are by nature, inter-disciplinary and multi-faceted. Obviously, their development involves close cooperation between area specialists in materials, electronics, mechanical engineering, chemistry, biology, various physics-based technologies and even economic services.

A sensor is the basic sensing element of a device. The development of sensors and transducers, their application and advisory services in the area of sensors and the relevant electronic devices is required to be considered as one of the major thrust areas in our country.

Since these devices were meant originally as an extension of human senses, it followed that they should be called "Sensors". To make the concept more clear, a sensor is defined as a non- elementary device, usually based on a transducer, capable of converting a physical non-electrical input quantity into an electrical output and of processing it, to provide control an output suitable for interfacing it to a process system.

Majority of the text books on the subject have followed a measurand-specific classification, like transducers for displacement, acceleration, force, temperature, pressure, strain etc. The present book has rather, ventured to classify transducers and sensors in a novel way, which may be called a principal variable-specific approach to their classification, like resistive transducers, inductive and capacitive transducers, ultrasonic transducers, piezoelectric transducers etc. Thus various ion selective electrodes including those used for pH measurement appear in chapter seven. Chapter eight deals with biosensors and transducers which are perhaps the latest in the line and have started coming up in a big way, during 1990s only. These have been developed as a result of close interaction between biotechnology and electronics. Lately, the application of fiber optics to biosensors has imparted a very quick and needful boost to these. Stated precisely, bio sensor is an analytical tool which uses biologically sensitive material to detect biological or chemical species directly without the need for complex processing of samples. Biosensors based on immobilized enzymes and micro-organisms have been exploited to a great extent, in such diverse areas as food and fermentation analysis, health care, environmental monitoring and even Defence. Being a recent craze, the subject of biosensors is bound to come up for study by degree engineering students if not to-day then surely very soon. Hence a small chapter on these.

Fiber optics has also permeated in the development of sensors and transducers of a wide variety and in view of their obvious advantages, their application areas are bound to grow. These have in fact, come up as a result of tremendous growth of optical fiber communication, with the easy availability of low loss optical fibers. Chapter nine therefore includes this category alongwith the traditional optical sensors whose application in radiation detection and non-contact temperature measurement is better known by the name radiation pyrometry.

Chapter ten on transducers giving digital output has dealt with mainly encoders both mechanical and optical type. These are very much the part of degree engineering syllabus of Indian Universities. These find use in a variety of applications, *e.g.* machine tools, gauging and Robotics, As a matter of fact, the first generation robots which were without any computing power, depended solely on data supplied by encoders to have a final control on their end arm positioning.

However, with a speedy advancement in robotics technology, and growth in the computing power of robots (so as to make them interact with their environment), need was felt of an altogether new class of sensors, *viz.* vision sensors, range finders and tactile sensors etc. This is a new and fast upcoming area. While the subject of robotics technology is finding its place in the syllabi of Indian Universities, the robotics sensors are still futuristic. A small chapter eleven has therefore been included so as to familiarize the students with the basic concepts of sensors for robotics and their geometry, with an eye on near future needs.

The last chapter is devoted to the subject of signal conditioning, signal generation and signal processing (both linear and non linear), using a variety of electronic devices. This has been given the generic title of transducer interfaces. This chapter combined with the first one, namely characteristics of signals and systems, makes many essential aspects of signals clear and lucid. In fact without these chapters, the whole treatment of sensors and transducers remains incomplete.

A good deal of information covered in this book is based largely on the exhaustive notes prepared by the author while pursuing a post-retirement teaching assignment at Sant Longowal Institute for Engineering & Technology, Longowal (Sangrur), Punjab, while the rest is the result of various discrete studies made during the years with CSIO, Chandigarh.

The continued and sustained efforts made by my son-in-law, Sanjay Sharma for composing and editing the manuscript on computer are gratefully acknowledged.

Thanks are also due to my daughter Abha for material help and my wife for moral support.

It is hoped that the book is well received by the students as well as the teaching faculty and that they get benefitted at the end.

M.G. JOSHI
Formerly, Scientist at Central Scientific
Instruments Organisation Chandigarh
GH-12/Sector-20
Panchkula

CENTRAL SCIENTIFIC INSTRUMENTS ORGANISATION
Sector 30-C, Chandigarh-160 020 (INDIA)

Dr. R. P. Bajpai
Ph. D, D.Sc, FIETE
DIRECTOR

Foreword

Shri M.G.Joshi has been associated with this organization as a scientist for over twenty five years. During this period, he has been editing various scientific publications with a good flair for writing and presentation. After retirement, he had been on a teaching assignment to degree engineering students and thus got ample exposure to the students community. Based on these, he has written a good book on the multi disciplinary subject of transducers which forms an essential aspect in the domain of instrumentation. Practically all types of transducers of current interest have been treated in a very lucid style and easy to understand language.

With an eye on futuristic advancement, various topics of the latest upcoming areas such as biosensors, robotic sensors, fiber optics based sensors have been included, for the interest of students. Side by side with this, the recent developments in piezoelectric polymers, poly vinylidiene fluoride (PVDF), practical applications of strain gauge transducers in road engineering, in wind tunnels for aerodynamic measurements and ultrasonic transducers for biomedical purposes have also been included and discussed in detail.

By and large, this book will be very comprehensive and interesting for engineering and science students.

(R.P.Bajpai)

B.N. Hi-Tech Machines (P) Ltd.
363, Industrial Area, Phase-II,
Panchkula (Haryana) 134 113

It was a great opportunity for me that I got a chance of reading this book and went through the contents therein. Creating a book like this is the work of great labour, time and dedication. I hope this book will certainly help engineers engaged in design, systems applications and industrial automations, Reasearch & Development etc. Enhancing their knowledge about the widest range of Transducers and its applications, thus enabling them to build ever-more advanced equipments with highest design & flexibility. It can also be of benefit to students of engineering faculties.

This book therefor serves a dual purpose, it fills a knowledge gap in the field of Transducer's technology and also offering a comprehensive text book on vast range of sensors as well Transducers etc.

B.N.GOSWAMI
MANAGING DIRECTOR

B.N.HI-TECH MACHINES PVT. LTD.
PANCHKULA (HARYANA)

CONTENTS

Chapter *Pages*

Chapter *Pages*

1

The Characteristics of Signals and Measurement Systems

1.1. CHARACTERISTICS OF SIGNALS

Signals could be either static or dynamic. The static signal implies the one that never changes and has therefore no information content but in practice this category includes those signals which change slowly, from one steady state to another. The dynamic signals on the other hand undergo quick changes and therefore require a measurement system that should be sufficiently fast acting so that the output faithfully follows the input.

The dynamic signals could be classified under periodic, transient and random signals. Each of these form can occur separate but more often the practical signals contain combination of two or all the three.

A periodic signal is one that repeats itself in a regular cyclic manner. So that knowing the behaviour of a signal over one cycle is enough to predict its form over subsequent periods of time. A transient signal, on the other hand, is isolated in time and does not repeat in a regular manner. In practice, a signal having duration short compared with the period of repetition, is also classified as a transient one.

Random signals are characterised by a non-periodic time variation. The only predictions that can be made regarding its future variations are of statistical nature and interest. Such random signals are always present to some extent in any measurement system and set an ultimate limit to the achievable sensitivity.

1.1.1. Periodic Signals

The simplest mathematical form of a periodic signal is the sinusoidal relationship of a wave given by

$$A = A_m \sin \omega t \qquad ...(1.1)$$

where, A_m is the peak amplitude,

$\omega = 2\pi f$, the angular frequency and the time period

$T = 2\pi/\omega$

Any complex periodic signal waveform can be analyzed into a series of sinusoidal signals and it is then much simpler to examine how the individual sinusoidal components are affected by the measurement system.

It can be shown that any periodic signal $f(t)$ of period T can be analyzed and split into a series of discrete sinusoidal waves given by :

$$f(t) = a_o/2 + \sum_{n=1}^{\infty} a_n \cos n\omega t + b_n \sin n\omega t \qquad ...(1.2)$$

In which the harmonic amplitude are given by

$$a_n = 2/T \int_0^T f(t) \cos n\omega t \, dt \quad \text{...(1.3)}$$

and

$$b_n = 2/T \int_0^T f(t) \sin n\omega t \, dt \quad \text{...(1.4)}$$

The series given by Eqn. (1.2), consists of a term $a_o/2$ equal to the average value of $f(t)$ over one period, plus the sum of a series of sine and cosine terms, the frequency components of which increase in integral steps from the fundamental frequency ω upto infinity. This series is called Fourier series.

Fig. 1.1 illustrates how three sets of waves marked 1, 3, 5 representing fundamental, third and fifth harmonic components can be seen to be beginning to approximate the shape of the square wave. An infinite number of harmonics will be required to be added up to give the exact shape of the square wave. Like this, through a judicious choice of magnitude and sign of the amplitude factors, the harmonic components add up to give a complex waveform.

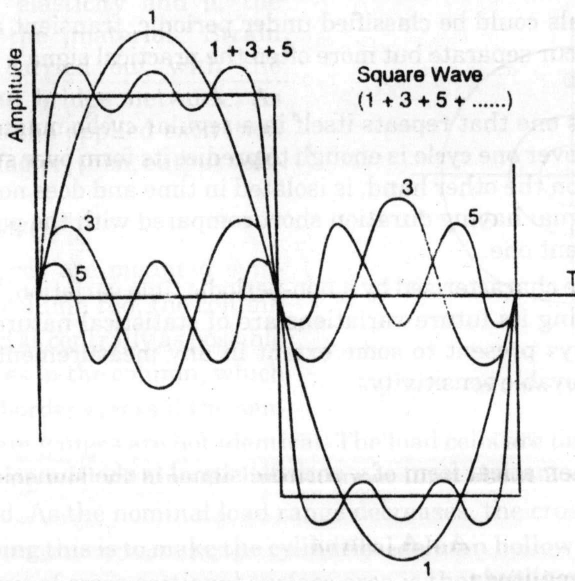

Fig. 1.1. Square wave approximation by Fourier Synthesis.

It can also be shown that if the signal is an odd function *i.e.* if $-f(t) = f(-t)$, the series will contain sine terms only. If the signal is even, *i.e.* $f(t) = f(-t)$, the series will contain cosine terms only.

Fig. 1.2 shows examples of even and odd periodic waveforms. A simple inspection of these waveforms enables one to decide if either set of terms a_n or b_n are zero.

The Eqns. (1.1) and (1.2) can also be rewritten by introducing a phase term ϕ_n as

$$f(t) = a_o/2 + \sum_{n=1}^{\infty} c_n \sin (n\omega t - \phi_n) \quad \text{...(1.5)}$$

where $c_n = \sqrt{a_n^2 + b_n^2}$ and $\phi_n = \tan^{-1} a_n / b_n$

This form is particularly valuable for examining the effect of frequency dependent phase shifts on the waveform.

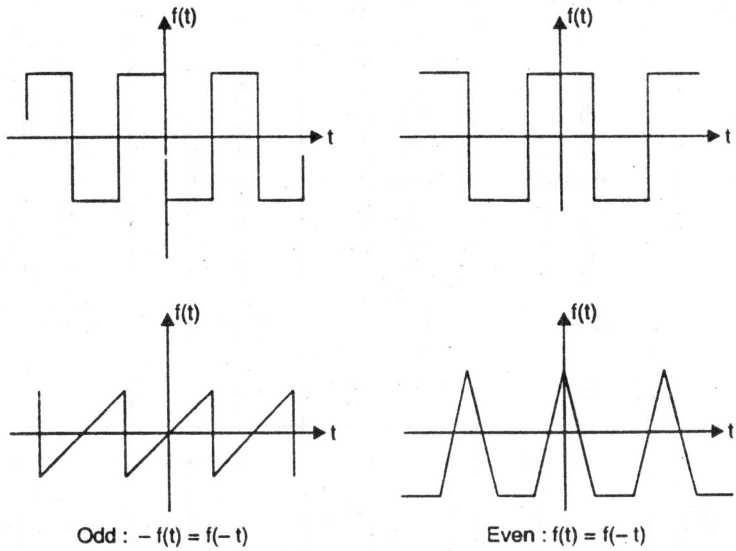

Fig. 1.2. Examples of even and odd periodic waveforms.

Evaluation of Eqn. (1.2) for simple waveforms, the harmonic amplitudes for smoothly varying curves are found to decrease rapidly with increasing harmonic number, while those changing not smoothly but abruptly, contain higher frequency harmonics. Thus, from Table 1.1, the harmonic amplitudes for the triangular wave shown at column 3 decrease relatively faster compared with those for the square wave. More abrupt and discontinuous the waveform, the larger is the number of terms to be included to obtain a reasonable approximation to the wave shape. Although the Fourier series, with an infinite number of terms included, can exactly represent any periodic waveform, in practice it is only desired to approximate the wave shape closely. Therefore, the number of terms to be included depends on the nature of the waveform and the accuracy desired.

The most important aspect, especially in the context of non-sinusoidal waveforms, is that of the frequency spectrum. Third column in Table 1.1 shows the frequency spectrum of certain periodic waveforms. Each vertical line indicates the amplitude of a specific harmonic. It is a characteristic of all perfectly periodic waves that they have a "line spectrum" of this form, which is in contrast with the "continuous spectrum", (wherein all frequencies are present), for random or transient signals. A waveform that is nearly periodic or a periodic waveform that contains a small amount of noise will exhibit a spectrum in which the lines are broadened and a small amplitude continuous spectrum is present. In medical practice, many physiological signals are of this form.

Analyzing the Periodic Signals. The determination of the frequency spectrum of waveforms, provides the necessary information for the proper design of a measuring system as well as about the system being measured in a form that can be more readily interpreted in terms of the physical processes involved. Various methods are applied in for spectral analysis of periodic waveforms and these depend on the frequency range and the nature of signal involved as also the form in which the signal is available (whether analog or digital).

Table 1.1. Fourier Series and Frequency Spectrum for Some Simple Periodic Waves

Wave Type	Waveform	Harmonic Amplitude (Frequency Spectrum)	Equation
Square Wave	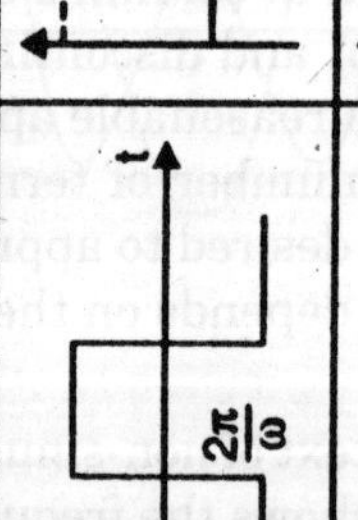	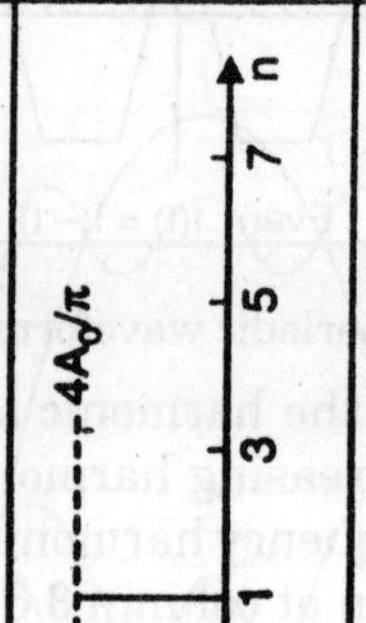	$f(t) = \frac{4A_0}{\pi}\left(\sin \omega t + \frac{1}{3}\sin 3\omega t + \frac{1}{5}\sin 5\omega t + \ldots\right)$
Saw Tooth Wave	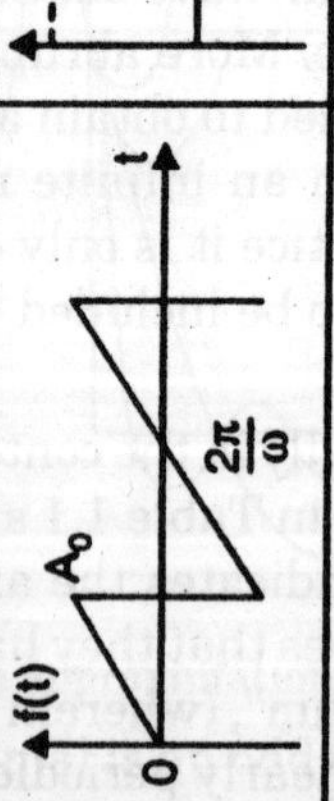	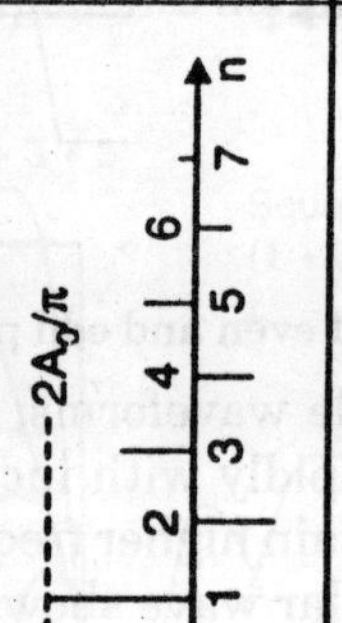	$f(t) = \frac{2A_0}{\pi}\left(\sin \omega t - \frac{1}{2}\sin 2\omega t + \frac{1}{3}\sin 3\omega t - \frac{1}{4}\sin 4\omega t + \ldots\right)$
Triangular Wave	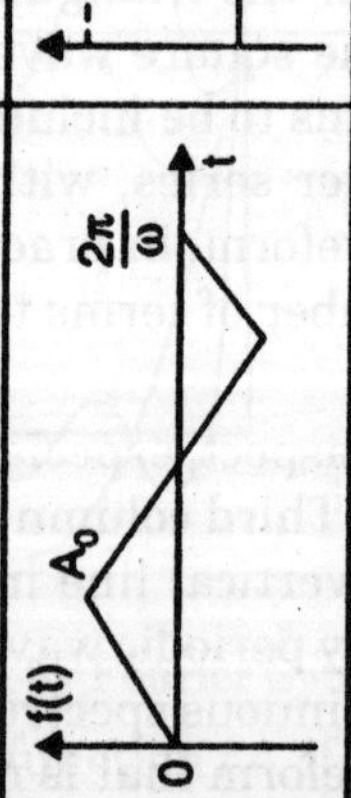	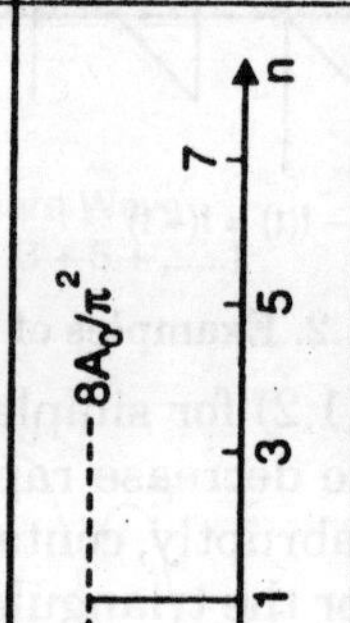	$f(t) = \frac{8A_0}{\pi}\left(\sin \omega t - \frac{1}{3^2}\sin 3\omega t + \frac{1}{5^2}\sin 5\omega t - \ldots\right)$
Half Rectified Wave	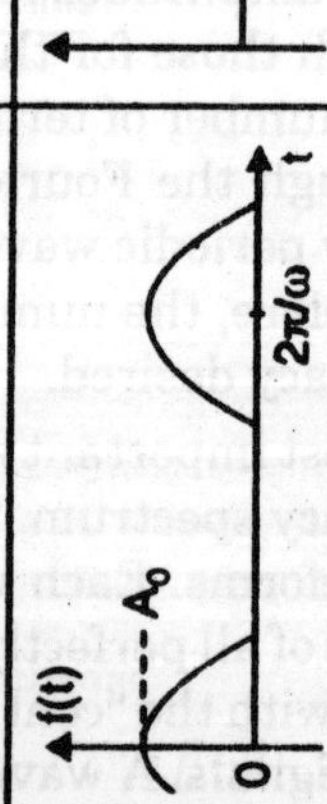	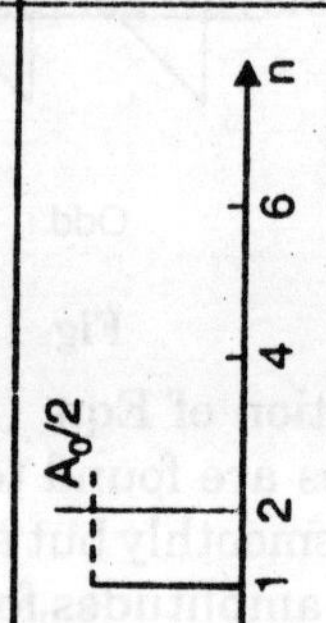	$f(t) = \frac{A_0}{2}\left(\frac{2}{\pi} + \cos \omega t - \frac{4}{3}\cos 2\omega t + \frac{4}{15}\cos 4\omega t - \frac{4}{35}\cos 6\omega t - \ldots\right)$

The use of a series of narrow band filters arranged to cover the frequency range of interest is the most straightforward but expensive method of a real time spectral analysis of both random

and periodic signals, available in analog form. One such method uses a series of 1/3 octave filters to cover a frequency range of 3.15 Hz to 80 KHz. A multiplexer and a digital detector are then employed to measure the output from each filter, in turn. Use of a large number of precision filters saves analysis time but adds to cost. Alternately, the incoming signal is mixed with a signal from a variable frequency oscillator. The sum or difference frequencies that result, are then passed through a single narrow band filter whose output is recorded. Various components of the signal spectrum are thus selected by changing the oscillator frequency and are measured. The system behaves as a tunable narrow band filter, the central frequency of which is determined by the oscillator. Hewlett-Packard have reported the use of a programmable local oscillator for making the whole operation automatic so that the central filter frequency is swept over the range of interest and the spectral characteristic, of the signal are displayed graphically.

Another method of analysis makes use of a Fourier transform of the auto correlation function, which in case of a periodic waveform of time period T can be defined as :

$$\Phi(t') = 1/T \int_0^T f(t)\, f(t - t')\, dt' \qquad \text{...(1.6)}$$

If $f(t)$ is given by Eqn. (1.1), then substituting Eqn. (1.1) into (1.6), we get :

$$\Phi(t') = a_o^2/4 + \frac{1}{2} \sum_{n=1}^{\infty} (a_n^2 + b_n^2) \cos n\omega t' \qquad \text{...(1.7)}$$

The Cosine Fourier transform of $\Phi(t')$, is by definition, given by

$$\Phi(n) = 1/T \int_0^T \Phi(t') \cos n\omega t'\, dt' \qquad \text{...(1.8)}$$

Substituting from Eqn. (1.7) into (1.8) :

$$\Phi(n) = (a_n^2 + b_n^2)/4 \qquad \text{...(1.9)}$$

Which is half the mean square value of the nth harmonic amplitude, Thus by finding the auto correlation function and taking its Fourier transforms enables the spectral amplitude to be determined.

A commercial system, (Princeton Applied Research Signal Correlation model 100 combined with Model 102 Fourier Analyzer), based on these two operations, enables us to analyze a waveform in less than a minute.

Digital techniques making use of either a general-purpose digital computer and an appropriate program or a machine designed specially for the type of computations involved provide a powerful means for both online and off line spectral analysis. Conversion of an analog input signal into digital form is usually accomplished by sampling the former at regular intervals and converting the sampled amplitude into a digital form. If the sampling frequency is greater than twice the highest harmonic frequency of practical interest, accurate results are achieved. (Refer section 2.2.13). The discrete Fourier transform, evaluated by means of a fast Fourier transform (FFT) algorithm, enables, one to derive the spectrum from the sampled information.

Still another method is through the use of optical techniques, which potentially offer an economical means for spectral analysis of the recorded data. The method makes use of the Fourier transformation properties of optical lenses. Using either a photographic transparency or an acoustical diffraction cell, one can introduce the signal waveform to be analyzed at the input focal plane of the system. An output is obtained whose optical intensity is a measure of the spectral amplitude and whose displacement is a measure of the harmonic number.

1.1.2. Transient Signals

Eqn. (1.2) can also be expressed in terms of complex numbers by making use of the relations :

$$\cos n\omega t = \frac{1}{2}\left(e^{jn\omega t} + e^{-jn\omega t}\right)$$

$$\sin n\omega t = \frac{1}{2}\left(e^{jn\omega t} - e^{-jn\omega t}\right) \quad ...(1.10)$$

This yields a compact form of Fourier series :

$$f(t) = \sum_{n=-\infty}^{+\infty} A_n e^{jn\omega t} \quad ...(1.11)$$

Where the complex amplitudes A_n are given by :

$$A_n = 1/T \int_0^T f(t)\, e^{-jn\omega t}\, dt \quad ...(1.12)$$

It can be shown that these amplitude terms are related to the real amplitude terms in Eqn. (1.2) by :

$$A_o = \frac{1}{2} a_o \quad A_n = \frac{1}{2}(a_n \pm jb_n), \ | \ n \ | > 0$$

where minus sign applies for positive values of n and the plus sign to negative n.

Just as we have developed the above complex periodic waves, likewise transients can also be synthesized from sinusoidal periodic waves. The major difference being that the spectrum is continuous instead of discrete and although the amplitude of any individual frequency is zero, the value of signal within a finite frequency range is non zero. For a periodic signal the amount of energy contained in the signal is composed of the energy content of a finite number of discrete frequencies that are harmonics of the fundamental. For a transient signal, if the energy content of a discrete frequency were non zero, then because of the infinite number of frequencies present, the energy content of the signal would be infinite. Hence, it follows that the amplitudes are zero, while the amount of signal within a bandwidth $\Delta\omega$ will be non-zero. This amount of signal is often called the spectral density, $g(\omega)$. It is analogous to the amplitude terms A_n contained in Eqn. (1.12) and is in general, a complex quantity defined by :

$$g(\omega) = \int_{-\infty}^{+\infty} f(t)\, e^{-j\omega t}\, dt \quad ...(1.13)$$

where $f(t)$ is the transient signal.

Just as for the periodic signal, a transient can be reconstructed by integrating the spectral density over all frequencies, that is

$$f(t) = \frac{1}{2\pi}\int_{-\infty}^{+\infty} g(\omega)\, e^{j\omega t}\, d\omega \quad ...(1.14)$$

This expression is analogous to the complex Fourier Series of a periodic wave as given by Eqn. (1.11).

The above Eqns. (1.13) and (1.14) form a Fourier transform pair. Eqn. (1.13) enables one to determine how the altered spectrum affects the signal shape. Therefore, Eqns. (1.13) and (1.14) form the basis of examining the effects of a signal conditioning system on the signal itself.

1.1.3. Noise

Noise is present to some extent in all measuring systems and in all signals sources. It is usual to distinguish between noise of a random nature and the noise caused by interference, which may or may not be correlated with the signal being measured. For example, one source of fundamental noise arises from statistical variations in the electron density in a conductor and will be present in all resistive elements. A frequent cause of interference noise is the electromagnetic or electrostatic interference arising from the presence of 60 Hz driven components. This latter type of noise can be eliminated by careful electrostatic and electromagnetic shielding. However, noise arising from basic physical phenomena usually sets a limit to the precision with which a given measurement can be performed. Generally, the noise content increases with the signal bandwidth, so that for dynamic signals requiring large bandwidths, this has to be taken care of, in order to ensure optimum performance of a measurement system.

If the sources of noise are not correlated with the signal, very significant reduction in noise content of periodic signals is possible by an averaging process. This is done by sampling the signal a large number of times over its period and storing the results in a digital computer. The process is repeated over successive cycles and the new amplitudes are added to the previous ones, thereby trying to get an average amplitude through sampling since random noise and uncorrelated interference signals have an average value of zero, by increasing the number of periods over which the signal is averaged, an improved signal-noise ratio results. In fact, the ratio can be improved by as much as the square root of the number of sampling cycles, provided the sync pulse responsible for initiating each pulse is properly locked to the signal being extracted.

The root mean square (RMS) value is a convenient measure of noise amplitude, since all negative amplitudes become positive by squaring and consequently, the RMS value is a non-zero quantity. However, besides noise amplitude, this method gives nothing about its frequency content. This can be specified by a graph of the spectral density as shown in Fig. 1.3. This graph displays three distinct regions : a low frequency zone in which the mean square noise value varies as $1/f$, a high frequency region displaying an increasing noise spectrum and an intermediate frequency zone in which the noise spectrum remains essentially flat. Most components and systems display a low frequency region in which the noise varies as $1/f^n$ (n is approximately unity). Typically, an amplifier may exhibit a noise spectrum in which intermediate frequency noise is dominant for frequencies less than a few KHz. This *I/f* noise is governed by thermal fluctuations in the electron density, giving rise to what is known as Johnson noise, which has a flat response. This noise is also commonly referred to as "white noise". As the frequency

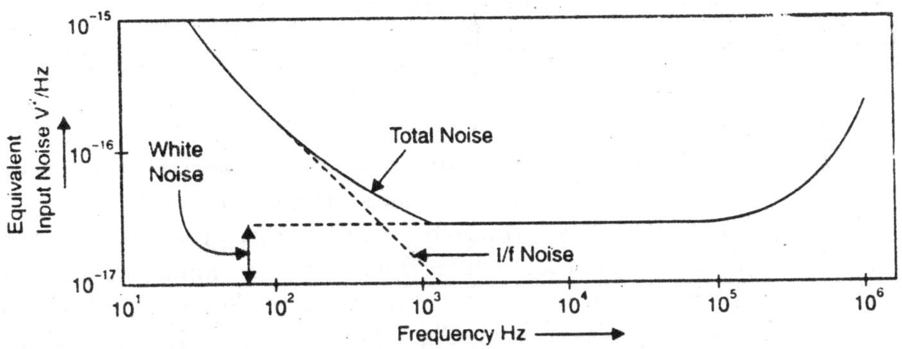

Fig. 1.3. Frequency spectrum of the equivalent of the mean square input voltage noise per cycle of bandwidth (Spectral Density) of a wide band signal amplifier.

approaches the cut off frequency of the amplifier, an increasing noise amplitude is generally observed. *I/f* noise is of interest in the amplification of low frequency signals or the signals in which the *DC* component must be preserved. By modulating the input signal so as to convert it to a higher frequency, the amplifier can be used in a region where the *I/f* component is less important. Once the signal has been amplified to such an extent that the noise generated by subsequent conditioning elements does not assume importance, the signal can be demodulated and the original but amplified signal can be recovered.

1.2. DYNAMIC CHARACTERISTICS OF MEASUREMENT SYSTEMS

The dynamic characteristics of a measurement system can be best realized by postulating a generalized mathematical model with due consideration to the dynamic input-output relationship. In other words, methods must be evolved for a quantitative evaluation of a system in terms of how it modifies the input signal, how it interacts with the signal source and how it is affected by an output load. For linear systems (in which input and output are linearly related), such quantitative evaluations can be best performed by specifying the input and output impedances, and the input-output transfer function. By restricting our analysis to the linear systems, the mathematical treatment becomes comparatively easier. Non linearity, hysteresis etc. have been discussed in Chapter 2, primarily as quasi static phenomena ; by superimposing these quasi static and dynamic characteristics, a rough measure of the overall system performance can be obtained.

1.2.1. Input-Output Transfer Function

Transfer function is defined as the ratio of the laplace transform of the output quantity to the laplace transform of the input quantity, when all initial conditions are zero. The generalized mathematical model for dynamic response of a system is described by an ordinary linear differential equation with constant coefficients :

$$a_n \frac{dx_o{}^n}{dt^n} + a_{n-1} \frac{dx_o{}^{n-1}}{dt^{n-1}} + \ldots\ldots + a_1 \frac{dx_o}{dt} + a_o x_o$$

$$= b_m \frac{dx_i{}^m}{dt^m} + b_{m-1} \frac{dx_i{}^{m-1}}{dt^{m-1}} + \ldots\ldots + b_1 \frac{dx_i}{dt} + b_o x_i \quad \ldots(1.15)$$

The operations transfer function of this equation can be written as :

$$\frac{x_o(s)}{x_i(s)} = \frac{b_m s^m + b_{m-1} s^{m-1} + \ldots\ldots + b_1 s + b_o}{a_n s^n \; a_{n-1} s^{n-1} \; a_1 s + a_o} \quad \ldots(1.16)$$

where $x_0(s)$ is the Laplace transform of the output and $x_i(s)$ the Laplace transform of input, S is the Laplace operator.

The operational transfer function can be used for calculating the response to transient input signals. However, the most important is the steady state response to the sinusoidal input. Then, x_i takes the form of $x_i = A_i \sin \omega t$. In all linear systems, after the dying out of the initial transients, the output quantity will be a sinusoidal wave of exactly the same angular frequency of the input. The amplitude may of course differ from that of the input and a phase shift may also get introduced. The frequency dependence of these quantities, amplitude and phase shift enables one to produce two graphs as shown in Fig. 1.4 (*a*), (*b*) and (*c*). Giving amplitude response and phase response respectively.

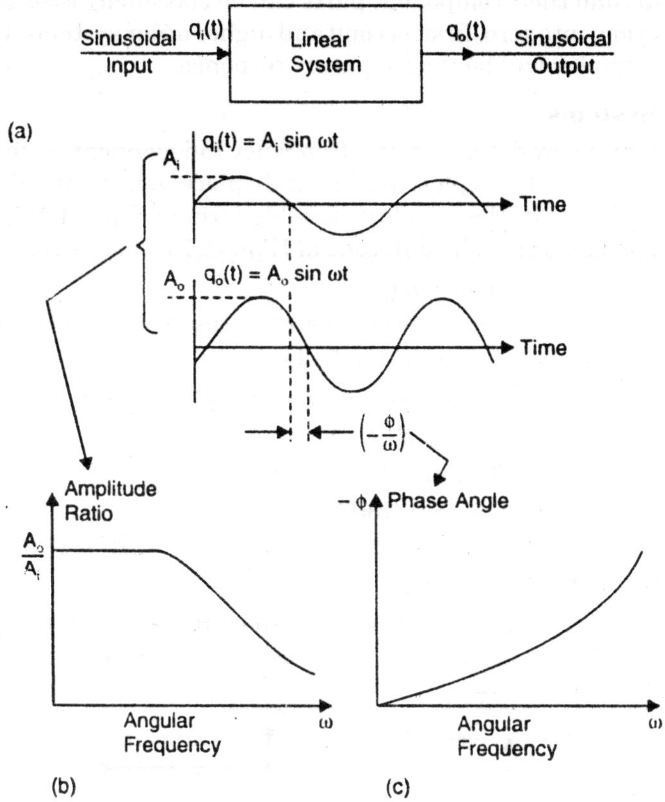

Fig. 1.4. Sinusoidal frequency response of a linear system curves
(*b*) and (*c*) are amplitude and phase response at various frequencies.

The frequency response of the system will consist of these curves sharing amplitude ratio and phase shift as a function of frequency.

The sinusoidal transfer function of a measurement system is obtained by the use of a complex number. This enables the transfer function to be expressed in a compact, and often a more convenient form because a complex number can represent magnitude as well as phase angle of a quantity.

The sinusoidal transfer function can therefore be obtained by substituting $j\omega$ for s whenever it appears in the operational transfer function expressed by Eqn. (1.16), as below :

$$\frac{x_o}{x_i}(j\omega) = \frac{b_m(j\omega)^m + b_{m-1}(j\omega)^{m-1} + \ldots\ldots + b_1(j\omega) + b_o}{a_n(j\omega)^n + a_{n-1}(j\omega)^{n-1} + \ldots\ldots + a_1(j\omega) + a_o} \quad \ldots(1.17)$$

For any given frequency ω, the term $\frac{x_0}{x_i}(j\omega)$ is a complex number.

This can also be expressed in polar form as :

$$\frac{x_o}{x_i}(j\omega) = \frac{|x_o|}{|x_i|} < \Phi = M < \Phi \quad \ldots(1.18)$$

where M represents the amplitude ratio of the output to input and the angle Φ represents the phase angle by which the output x_o leads the input x_i.

Linear systems and their component parts can be classified, according to the complexity of their transfer function, into zero, first, second and higher order systems. In general, the higher the order, the more complex will be the frequency response.

1.2.2. Zero Order Systems

A zero order system, by definition, is a frequency independent system in which the input and output are proportional. Examples, are linear displacement or angular displacement type potentiometric transducers. In the simplest possible form of Eqn. (1.15), all a's and b's other than a_o and b_o are assumed zero. The differential Eqn. (1.15) then becomes a simple equation

$$a_o x_o = b_o x_i \qquad ...(1.19)$$

The static sensitivity of the zero order system k therefore, can be expressed as :

$$x_o = (b_o/a_o)x_i = kx_i \qquad ...(1.20)$$

In the practical example of linear potentiometer shown at Fig. 1.5 (a), the output voltage

$$V_o = (R_1/R_2)V_i$$

and in case of an angular potentiometer shown at Fig. 1.5 (b),

$$V_o = (\theta/180)V_i$$

The transfer function in either case will be given by : $V_o/R_1 = V_i/R_2$ and $V_o/\theta = V_i/180$ respectively.

Since, it is never possible to realize a pure resistance element, devoid of any inductance or capacitance (depending on winding), the zero order system represented above may not strictly

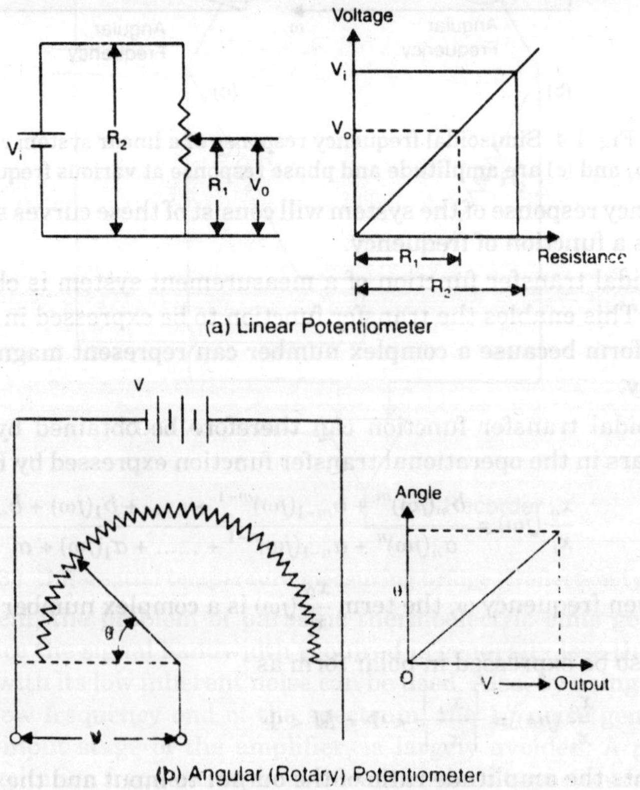

(a) Linear Potentiometer

(b) Angular (Rotary) Potentiometer

Fig. 1.5

be a zero order one. Also, in the case of the rotary or angular potentiometer, the inductance and capacitance effects could be neglected if the changes in angle θ are sufficiently slow.

Thus, it can be concluded that the system may be of an order higher than zero, the frequencies involved are considered to be of zero order. In fact, many higher order systems could be approximated by a zero order system if changes do not occur too rapidly or the frequencies involved are not too high.

1.2.3. First Order Systems

Unlike zero order systems, the first order systems contain a single energy storage element and have a transfer function that can be expressed in the form

$$F(j\omega) = K/(1 + j\omega T)$$

where T is the time constant of the system and K is constant. From this one can find the magnitude and phase angle of the output-input ratio, as :

$$| F(j\omega) | = K/\sqrt{(1 + \omega^2 T^2)} \text{ and } \phi = \tan^{-1} (- \omega T)$$

Alternatively, in the generalized differential Eqn. (1.15), if second and higher order terms are assumed to be zero, we get the equation of first order :

$$a_1 \frac{dx_o}{dt} + a_o x_o = b_o x_i \qquad ...(1.21)$$

or

$$\frac{a_1}{a_o} \frac{dx_o}{dt} + x_o = \frac{b_o x_i}{a_o} \qquad ...(1.22)$$

Putting $b_o/a_o = K$ (static sensitivity) and $a_1/a_o = T$ (Time constant) and taking laplace transform with zero initial conditions,

$$\text{The transfer function} \quad \frac{x_o}{x_i}(S) = \frac{K}{TS + 1} \qquad ...(1.23)$$

or

$$F(j\omega) = K/(1 + j\omega T)$$

where laplace operator S has been replaced by the complex variable $j\omega$.

A mercury-in-glass thermometer is a typical example of the first order system.

When such a thermometer is dipped into a mass of fluid contained in a vessel, the heat balance equation will be given by :

$$h_t a_b (\theta_i - \theta_f)\, dt = \rho\, c v_b\, d\theta_f \qquad ...(1.24)$$

where h_t = Heat transfer coefficient across the bulb wall in calories/cm^2 °C sec
θ_i = Temperature of fluid in °C
θ_f = Temperature rise in thermometer mercury in °C
a_b = Area of heat transfer of thermometer bulb in cm^2
dt = Small interval of time
ρ = Density of mercury, (13.6)
c = Specific heat of mercury
v_b = volume of bulb in cc
$d\theta_f$ = temperature rise in a small time dt

From Eqn. (1.23), we get,

$$\frac{\rho c\, v_b}{h_t\, a_b} \cdot \frac{d\theta_f}{dt} + \theta_f = \theta_i \qquad ...(1.25)$$

where θ_i is the process condition determined by the thermometer column y expressed with a conversion factor K where

$$K = \alpha_{ex}\, v_b/a_c$$

α_{ex} = Differential coefficient of volume expansion of the bulb and fluid and a_c = capillary bore area of the thermometer in cm^2

$\rho c\, v_b$ = Thermal capacity and $1/h_t\, a_b$ = Thermal resistance

The Eqn. (1.24) then reduces to :

$$\frac{\rho c\, v_b}{h_t\, a_b} \cdot \frac{dy}{dt} + y = K\theta_i = Kx$$

where $\frac{\rho c\, v_b}{h_t\, a_b} = T$ (time constant) and $K = \alpha_{ex}\, v_b/a_c$

For a small response time, ρ, c, v_b are too small and h_t, a_b are large.

First Order System Step Response. When a step is applied to a first order system, initially in equilibrium and the input quantity is measured instantaneously by an amount x_s then Eqn. (1.23) becomes :

$$(TS + 1)x_o = \frac{Kx_s}{S} \qquad ...(1.26)$$

A typical example is a thermocouple in air suddenly immersed in boiling water, resulting in a step input change in the junction temperature. The step input can be represented mathematically by the relationship in time domain as :

$$x_i = 0 \text{ for } t = 0$$

And $x_i = x_s$ for $t \geq 0$

In laplace transform notation, $x_i(S) = x_s/S$

therefore Eqn. (1.26) becomes : $x_o(S) = \frac{K}{(TS + 1)} \cdot \frac{x_s}{S}$

or

$$x_o(S) = \left[\frac{1}{S} - \frac{1}{S + 1/T}\right] Kx_s \qquad ...(1.27)$$

The inverse Laplace transform of this becomes :

$$x_o(t) = (1 - e^{-t/T})\, Kx_s \qquad ...(1.28)$$

This shows that the speed of response depends only on the value of time constant T and the response is faster if T is smaller. Therefore, for a first order system, minimizing the magnitude of T helps to improve the dynamic response.

Rewriting Eqn. (1.28), $x_o/Kx_s = (1 - e^{-t/T})$ ...(1.29)

Plotting x_o/Kx_s against t/T, a response shown in Fig. 1.6 is obtained. One of the important characteristics of such an exponential response curve is that at $t = T$, $x_o = 0.632\, Kx_s$ or in first order systems, when $t = T$, the response reaches 63.2% of its steady state value x_s.

The first order system approaches ideal conditions when $T \to 0$. For any other value of T, the measurements will be accurate below a certain value of ω only. For a pure sinusoidal input, the amplitude and phase shift can be computed by simple calculations. If the input is complex, *i.e.* a combination of several sine waves of different frequencies, the transfer function at each frequency is to be computed and then interpreted.

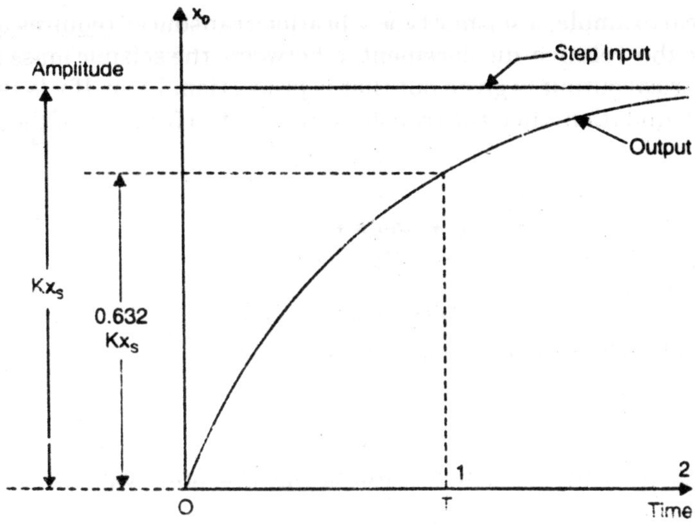

Fig. 1.6. Response of a first order system to a step input.

Example. Suppose a unit step input given to a first order system induces a response that rises to 95% of the final value of the impressed step value in 500 ms, obtain the measuring system parameters. If the time constant is halved, calculate the % improvement in speed of response.

From Eqn. (1.29), $y = Kx_o(1 - e^{-t/T})$

or $0.95 = 1 - e^{-0.5/T}$ or $e^{-0.5/T} = 0.05$

or $-0.5/T = \log_e 0.05 = -3.54$ or $T = 0.5/3.54 = 0.141$ sec or 141 ms

If this is halved *i.e.* 70 ms

$e^{-t/0.0835=0.05}$ which gives $t = 0.2478$ sec or 247.8 ms

therefore % improvement in speed of response (an index for time lag)

$= (500 - 247.8)/500 = 50\%$ approx.

1.2.4. Second Order System

Although various practical examples exist of the first order systems, it must be borne in mind that most of the systems are of a second or higher order.

A second order system can be defined as :

$$a_2 \frac{d^2x_o}{dt^2} + a_1 \frac{dx_o}{dt} + a_o x_o = b_o x_i$$

Taking laplace transform of this equation,

$$\frac{x_o}{x_i}(S) = \frac{K}{s^2/\omega_n{}^2 + (2z/\omega_n)S + 1} \quad ...(1.30)$$

where $\omega_n = (a_o/a_2)^{1/2}$, the undamped natural frequency in radians/sec and

$z = a_1/2\,(a_o a_2)^{1/2}$, the damping ratio

And $K = b_o/a_o$ the static sensitivity.

The typical examples of a second order system are spring balance and spring-mass systems employed for acceleration and force measurements.

Taking a typical example, a seismic type vibration transducer requires a single degree of freedom and senses the relative displacement x between the seismic mass and the platform [Fig. 1.7 (*a*)], an electrical output is given out that is proportional to x. If y is the corresponding displacement in the foundation, then the transfer function as a function of the Laplace variable s will become :

$$\frac{x}{s^2 y} = \frac{1}{s^2 + 2z\omega_n s + \omega_n^2} \quad ...(1.31)$$

which can be rewritten as :

$$\frac{x}{y} = \frac{s^2/\omega_n^2}{s^2/\omega_n^2 + 2zs/\omega_n + 1} \quad ...(1.32)$$

For a sinusoidal excitation at a frequency ω, $s = j\omega$, so

$$\frac{x}{y} = \frac{-(\omega/\omega_n)^2}{1 - (\omega/\omega_n)^2 + 2jz(\omega/\omega_n)} \quad ...(1.33)$$

Putting the ratio of excitation frequency to undamped natural frequency $\omega/\omega_n = c$, Eqn. (1.33) becomes

$$\frac{x}{y} = \frac{-c^2 e^{j\theta}}{[(1-c^2)^2 + (2zc)^2]^{1/2}} \quad ...(1.34)$$

where $\theta = -\tan^{-1}[2zc/(1-c^2)]$ ...(1.35)

Fig. 1.7 shows the displacement response curves of the seismic transducer described above.

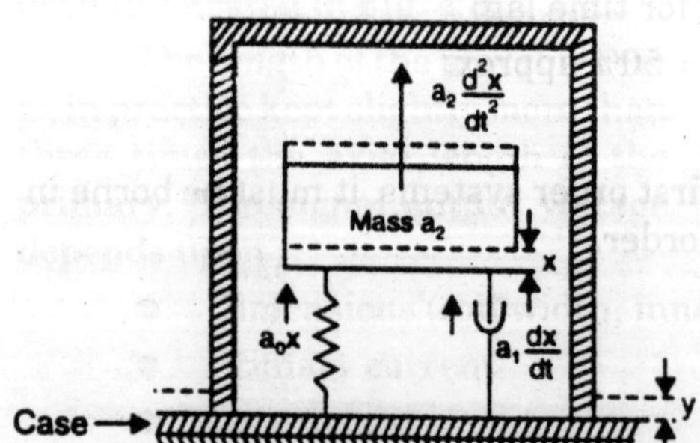

(*a*) Seismic spring mass system (Single degree of freedom)

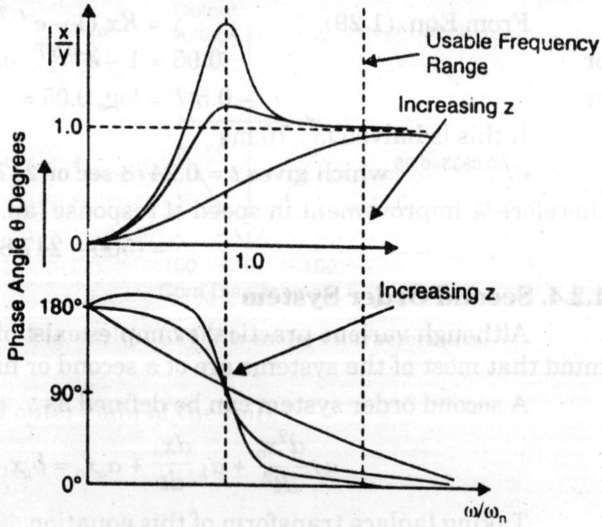

(*b*) Displacement response curves of seismic transducer.

Fig. 1.7

For large values of c (input frequencies much higher than the natural frequencies), the displacement of mass relative to the foundation frame become identical to the input displacement. This is evident from all the curves converging at $x/y = 1$. The phase angle curves indicate a noticeable phase shift at lower frequencies, which in turn, introduces distortion. With low frequency measurements, the natural frequency ω_n should be as low as possible which calls for a larger mass of the seismic element. This sets a limitation to the measurement of seismic displacement at very low frequencies. However, there is no high frequency limit for the displacement measurement.

1.2.4.1. Step Response

For a step input, the Eqn. (1.30) can be re-written as :

$$\frac{x_o}{Kx_s} = \frac{\omega_n^2}{s(s^2 + 2z\omega_n S + \omega_n^2)} \qquad \text{...(1.36)}$$

(The Laplace transform of step input is $1/s$)

The solution of Eqn. (1.31) is slightly complex but practical situations can be identified, depending upon the magnitude of z. The roots of $(s^2 + 2z\omega_n S + \omega_n^2)$ become real and separate for damping ratio greater than unity or $z > 1$, a case of over damping. The case $z = 1$ is that of critically damped system and this gives rise to real and repeated roots and the under damped system $z < 1$ gives a complex conjugate pair of roots.

Therefore,

(*a*) over damped system, $z > 1$,

$$\frac{x_o}{Kx_s}(t) = -\frac{z + (z^2-1)^{1/2} \exp.[-z + (z^2-1)^{1/2}\omega_n t]}{2(z^2-1)^{1/2}} + \frac{z - (z^2-1)^{1/2} \exp.[-z - (z^2-1)^{1/2}\omega_n t + 1]}{2(z^2-1)^{1/2}} \qquad \text{...(1.37)}$$

(*b*) critically damped, $z = 1$

$$\frac{x_o}{Kx_s}(t) = 1 - (1 + \omega_n t)\exp.(-\omega_n t) \qquad \text{...(1.38)}$$

(*c*) under damped, $z < 1$

$$\frac{x_o}{Kx_s}(t) = 1 - \exp.\frac{(-z\omega_n t)}{(1-z^2)^{1/2}} \sin(1-z^2)^{1/2}\omega_n + \varphi \qquad \text{...(1.39)}$$

where $\varphi = \sin^{-1}(1-z^2)^{1/2}$

Fig. 1.8 (*a*) shows a plot of x_o/Kx_s vs $\omega_n t$ for a step input. However, it must be borne in mind that it is the transient part of the differential equation for second order which has offered solutions mentioned above. So far as the solution in steady state is concerned, it will be $x_o = Kx_s$ under all situations. In a second order system, ω_n is a direct index of the speed of response just as it the time lag in case of the first order system. If angular frequency ω_n is doubled and t is halved, the normalized amplitude will attain the same value. An increased damping reduces oscillations but slows down the response and consequently the final value is reached after a longer duration. Another measure for the speed of response is called "Rise time" which is defined as the time required for output to rise from 10% to 90% of its final steady state value.

For a second order under damped system, the time taken by the output to reach the peak overshoot is also taken as a good measure of the speed of response. A plot of the output $x_o(t)$ against time t in case of an under damped system is shown at Fig. 1.8 (*b*). It exhibits an overshoot. The first overshoot occurs (or the first peak value) at time $t_{max} = \pi/\omega_n(1-z^2)^{1/2}$ and the magnitude of output at this overshoot is given by :

$x_{o\,max} = 1 + \exp.[-\pi z/(1-z^2)^{1/2}]$ which is a function of z only.

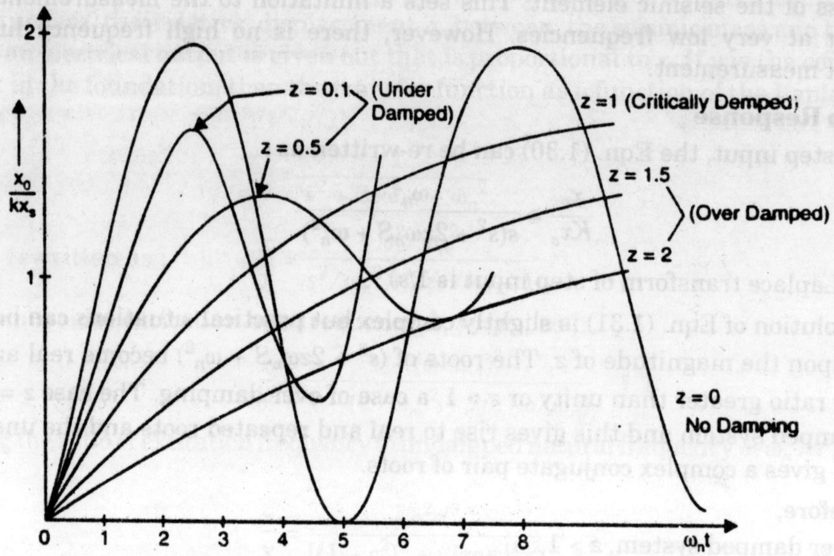

(*a*) Response of a second order system to a step input.

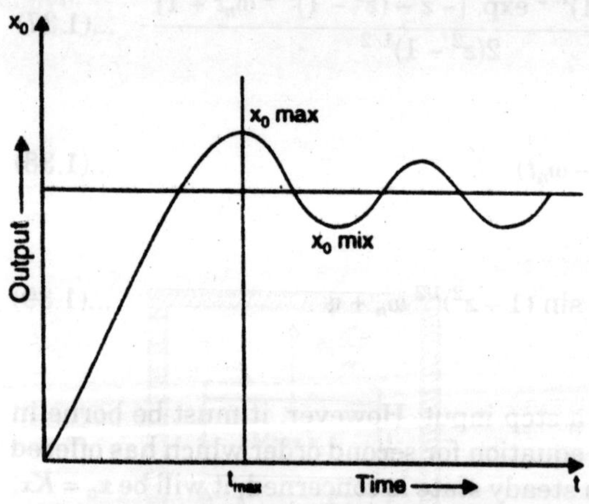

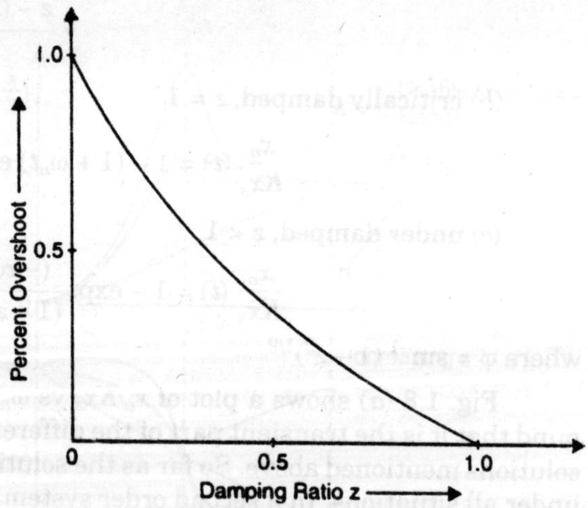

(*b*) A plot of time *Vs* Output for an under damped system.

(*c*) A plot of percent overshoot with damping ratio.

Fig. 1.8

1.2.4.2. Frequency Response

The transfer function of a second order system at steady state, for a sinusoidal input will be given by .

$$\frac{x_o}{x_i}(j\omega) = \frac{K}{(j\omega/\omega_n)^2 + 2zj\omega/\omega_n + 1} \quad ...(1.40)$$

This can be rewritten as :

$$\frac{x_o}{Kx_i}(j\omega) = \left[\frac{1}{[1-(\omega/\omega_n)^2]^2 + 4z^2\omega^2/\omega_n{}^2}\right]^{1/2} < \varphi \quad \text{...(1.41)}$$

$$\text{where } \varphi = -\tan^{-1}\frac{2z\omega/\omega_n}{(\omega/\omega_n)^2 - 1} \quad \text{...(1.42)}$$

The frequency and phase response are shown at Fig. 1.9 (*a*) and (*b*):

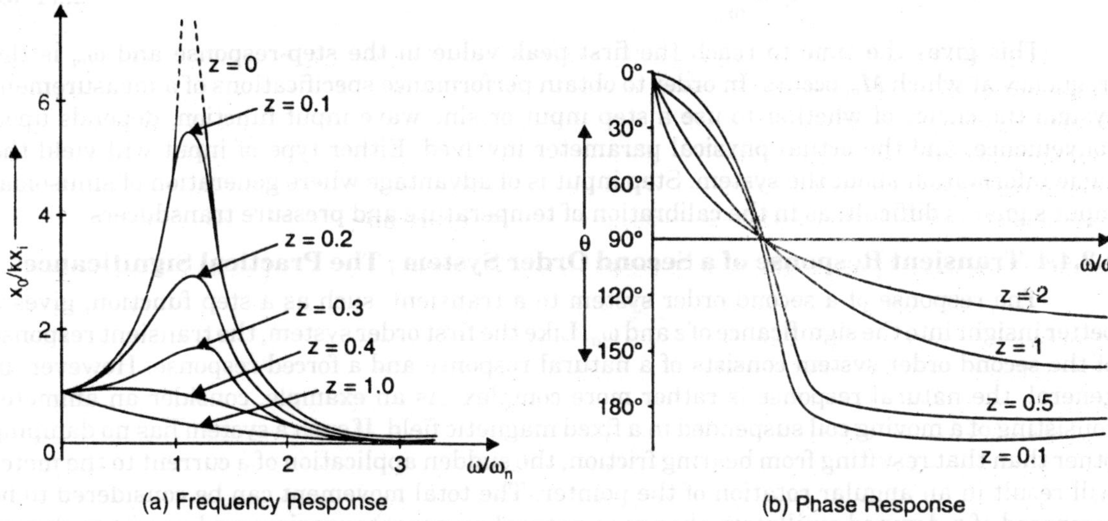

Fig. 1.9. Frequency and phase response of a second order sysem.

1.2.4.3. Dynamic Performance

Two basic factors, namely the speed of response and relative stability (degree of damping) are given to specify the dynamic performance of a measurement system. Most dynamic performance terms are usually interpreted in terms of the definitions of a linear second order system. These include the damped natural frequency and the damping ratio. Also, the dynamic performance may be defined either in time domain or frequency domain. The transient response to a step input is a common means of specifying the dynamic performance in the time domain. The terms that are used in specifying the speed of response are the rise time t_r which has already been defined earlier and the settling time t_s which is the time required for the output to remain within a certain percentage (typically 5%) of the final output. For a first order system, the time constant which is the time required for the output to reach 63% of its final value is used to define the speed of response.

Relative stability is generally characterised by the damping ratio. The term usually used to define relative stability in time domain, is the overshoot, or the maximum value to which the output reaches over the equilibrium value.

In the frequency domain, the specification is in terms of the system bandwidth, indicating the speed of response and the resonance peak M_m indicating the relative stability of the system. For a second order system the frequency ω_m at which M_m occurs is obtained by differentiating the term indicating the magnitude of the output in Eqn. (1.36), with respect to ω and equating to zero, or $$\omega_m = \omega_n(1 - 2z^2)^{1/2} \text{ (for } z \le 0.707) \quad ...(1.43)$$

where ω_n is the undamped natural frequency. The magnitude of the peak value is given by :

$$M_m = \frac{1}{2z(1 - z^2)^{1/2}} \quad ...(1.44)$$

For a system of second order, the response in the time domain is related to the frequency domain by the relationship :

$$t_p = \frac{\pi}{\omega_m} \quad ...(1.45)$$

This gives the time to reach the first peak value in the step-response and ω_m is the frequency at which M_m occurs. In order to obtain performance specifications of a measurement system the choice of whether to use a step input or sine wave input function, depends upon convenience, and the actual physical parameter involved. Either type of input will yield the same information about the system. Step input is of advantage where generation of sinusoidal input signal is difficult, as in the calibration of temperature and pressure transducers.

1.2.4.4. Transient Response of a Second Order System ; The Practical Significance

The response of a second order system to a transient, such as a step function, gives a better insight into the significance of z and ω_n. Like the first order system, the transient response of the second order system consists of a natural response and a forced response. However, in general, the natural response is rather more complex. As an example, consider an ammeter consisting of a moving coil suspended in a fixed magnetic field. If such a system has no damping other than that resulting from bearing friction, the sudden application of a current to the meter will result in an angular rotation of the pointer. The total movement can be considered to be composed of a damped oscillatory change (a natural response) superimposed on a step change (forced response) as illustrated in Fig. 1.10. The time taken for the deflection to settle to its final value is evidently greatly increased by the poorly damped oscillation. It can be reduced by

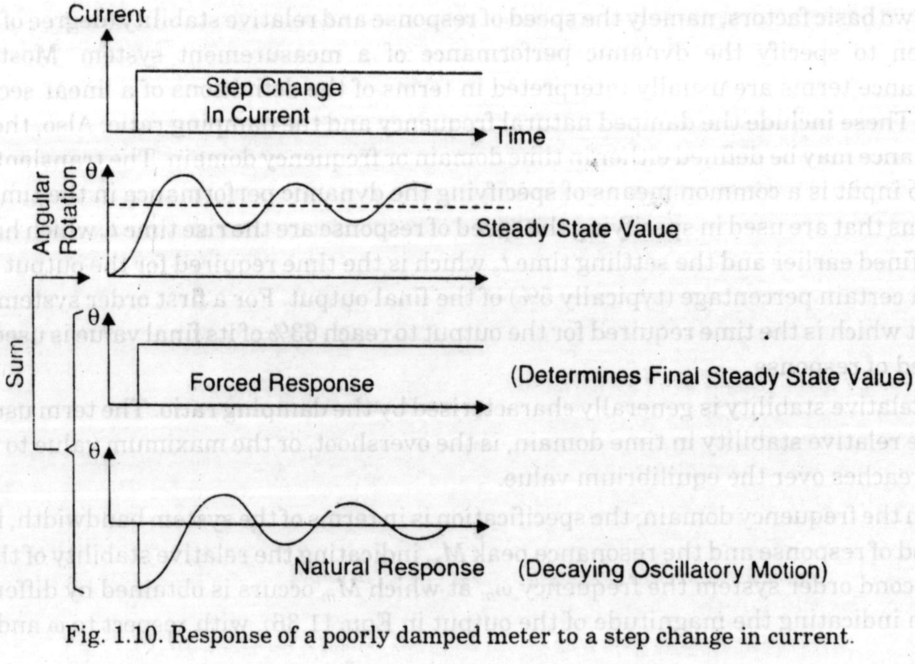

Fig. 1.10. Response of a poorly damped meter to a step change in current.

increasing the system damping through the addition of a resistance across the meter (resistive damping) or by using eddy currents induced in a metal plate (eddy current damping).

EXERCISES

1. **A linear, second order system, with a single degree of freedom has a mass of 10 g (10^{-4} kg) and a stiffness of 1000 N/m. Calculate the natural frequency and the damping ratio required to just prevent overshoot in response to a step input of force.**
2. **A temperature sensor having a first order response with a 1 sec time constant is given a step input of 50°C rising from 0°C. Calculate the temperature indicated 0.8 sec. after the input is applied. Plot the temperature response characteristics upto 2 seconds.**
3. **For a second order, single degree of freedom accelerometer, plot the maximum ratio of the input frequency to the natural frequency as a function of its damping ratio.**
4. **A second order instrument is required to measure the input signals upto 100 Hz with an amplitude of inaccuracy of less than 1%. Determine the natural frequency requirements and also the phase shift in each case at the frequency of 50875 Hz. The damping ratio and 0.3 and 0.6 respectively.**

BIBLIOGRAPHY

1. CS Rangan, GR Sarma and VSV Mani, *Instrumentation : Devices and Systems*, Tata-Mcgraw Hill, 1983.
2. E.O. Doebelin, *Measurement Systems ; Application and Design*, Mcgraw Hill New York, 1975.
3. J. Davies, *Introduction to Dynamic Analysis and Automatic Control* (Response of overall systems), John Wiley, New York, 1965.
4. Richard SC Cobbold, *Transducers for Biomedical Measurements : Principles and Applications*, John Wiley, New York, 1974.
5. DVS Murthy, *Transducers for Instrumentation Systems*, Prentice Hall of India Ltd., 1995.
6. A Popoulis, *The Fourier Integral and its Application*, Mcgraw Hill, New York, 1962.

2

Measurements and General Properties of Transducers

A detailed knowledge of measurement process and interaction of the measurement process with the parameters being measured are necessary for understanding the factors influencing the accuracy of measurement. Measurement provides us with a means of describing a natural phenomenon in quantitative terms. It involves a precise measurement and recording of physical and chemical quantity. Recent advances in electronics, physical and material sciences as well as technology have helped in the development of a number of sophisticated and high precision instruments and devices that can take care of a variety of measurement problems.

Industry has been using instrumentation in one form or the other for the last seven decades. In the intervening years the variety of variable parameters requiring measurement has increased manifold, with the result that the measuring techniques have also been improving considerably to meet the exacting demands of scientists, engineers, and technologists.

2.1. A generalised measurement system as shown in Fig 2.1 comprises the following elements :

(*i*) A transducer, which converts a measured quantity into a usable electrical output.

(*ii*) A signal conditioner, which transforms the transducer output into an electrical signal suitable for display, record or control. It may cover linearising a non-linear output, amplify a relatively weak transducer signal or even remove noise from the signal received from the transducer.

(*iii*) The display or readout devices or a recorder/printer unit. And

(*iv*) An electrical power supply, which provides necessary excitation to the transducer, signal conditioner and display devices. The transducer may be active or passive, depending upon whether or not it requires an external power source. A calibration signal source (*e.g.* a standard pressure or temperature) completes the system measurement.

(*v*) So far it has been assumed that the transducer produces an output signal that is directly related to the input and after conditioning, produces a proportional output. In practice this may not be quite true. The modern measuring systems therefore also contain a feedback. In case of a deviation of the signal conditioner output from that of transducer, a feedback difference causes the system to adjust itself to a null balance condition. The quantity adjusted by the feedback is then a correct measure of the quantity being measured. Feedback is sometimes also used to improve the system linearity and stability and to improve the input characteristics of the transducer itself.

(*vi*) A transducer is defined as a device which, when actuated by one form of energy, is capable of converting it to another form of energy. Although mechanical transducers, like dial

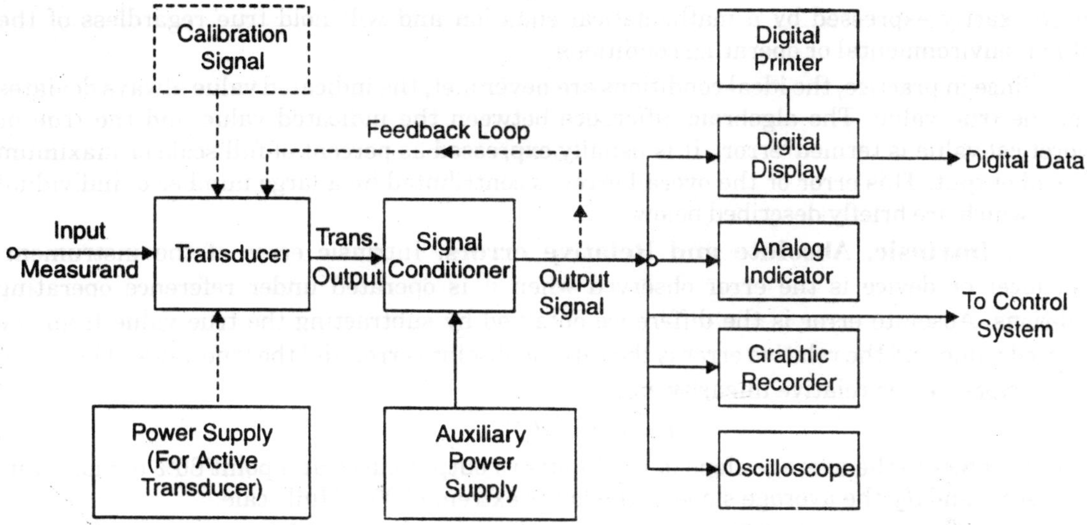

Fig. 2.1. A generalised measurement system (Broken lines are paths by choice).

gauge or an aneroid barometer have been in use for a long time, these do possess high accuracy, ruggedness, low cost, capable of operating without any external power but have a poor frequency response, need large forces to overcome mechanical friction and are incompatible to remote control or indication. Due to these limitations, we are these days, more concerned with the electrical transducers. These transform the measured physical quantity into an electrical voltage or current proportional to the measurand.

(vii) The input *vs* output energy relationship is a function that becomes reproducible. This reproducibility therefore becomes a prime criterion of a measurement system. Likewise, output-input relationship and its behaviour over time can be predicted with a definite accuracy, sensitivity and response, within specified environmental conditions.

2.2. BASIC CHARACTERISTICS OF MEASURING DEVICES

2.2.1. Accuracy

Accuracy can be defined as the closeness with which the reading approaches, an accepted standard value or the true value. Accuracy is determined by calibration under certain operating conditions and is expressed as within plus or minus a certain specified amount or a percentage at a certain point of scale or between specified points of scale.

2.2.2. Precision

By precision is meant the closeness with which the individual measurements are distributed about their value, meaning thereby, the degree of agreement of a set of measurements among themselves. It combines the uncertainity due to both random differences in results in a number of measurements and the smallest readable increment in scale, given as the deviation of mean value. Obviously, precision does not give a simultaneous guarantee on accuracy.

2.2.3. Error

For every transducer, there is an ideal or theoretical value of the output. This assumes an ideal design, very appropriate choice of materials and an ideal workmanship. This ideal value can be exactly expressed by a mathematical equation and will hold true regardless of the ambient environmental or operating conditions.

Since in practice, the ideal conditions are never met, the indicated value always deviates from the true value. The algebraic difference between the indicated value and the true or theoretical value is termed 'error'. It is usually expressed as percent of full scale or maximum value of output. This error or the overall error is contributed by a large number of individual errors, which are briefly described below :

(*a*) **Intrinsic, Absolute and Relative errors.** Intrinsic error of the instrument, transducer or device is the error observed when it is operated under reference operating conditions. Absolute error is the difference obtained by subtracting the true value from the observed value and the relative error is the ratio of absolute error and the true value. This could also be expressed as relative linearity error,

$$d = d_1 - d_2/d_1,$$

where d_1 refers to the average slope of the inputs *vs* output curve at a point 80% of maximum of full scale and d_2, the average slope at the lower extreme 10% of full scale.

(*b*) **Uncertainity and Random error.** These are observed when repeated measurements of the same quantity give differing values (due to lack of repeatability). These are caused by friction, hysteresis, noise etc. Other contributing factors are random changes in input signal, combined with noise and drift in the signal conditioner. The uncertainity is expressed as the average deviation, probable error or statistical deviation.

(*c*) **Interference Error.** The unwanted disturbances superimposed on low level input signal due to noise, line pickup, ripple, switching transients and line transients are termed as interference errors.

These are created by noise from electrical machines, magnetic fields, thermal sources, atmospheric interferences, electrostatic noise, noise in cables, arcing contacts in switches and relays etc. This error is reduced by isolation or frequency discrimination. The isolation may be electrical, electromagnetic electrostatic.

(*d*) **Application Errors.** Such errors are caused by incorrect or improper and faulty installation, especially if the device/transducer is used beyond the operating range, at high temperatures and vibration. Poor impedance matching with the signal conditioners could also result in application or installation errors.

(*e*) **Zero Drift.** This signifies the deviation observed in output from the initial values over a period of time—all other measurement conditions remaining constant. The reasons could be change in the value of components due to variation in ambient conditions or ageing with time.

(*f*) **Error due to sensitivity change.** The error due to drift at zero or full scale settings may often be completely random in nature and their correction may be difficult. Maximum difference usually occurs at the start and gets reduced after the warm up period. Changes in sensitivity are caused by temperature variations and applied voltage fluctuations. Errors of this type could be reduced by incorporating temperature compensation and voltage regulation by using a balanced differential amplifier or a stabilised chopper amplifier. The error caused by uncertainity in readings can be minimised by taking a large number of readings with static calibration while keeping the input constant. A random error of this type often follows a Gaussian distribution.

2.2.4. Linearity

The transducers designed to show a linear output *vs* input relationship are most common and are preferred. The linearity is expressed by the equation $y = mx + c$ where y and x indicate the output and input respectively, m is the slope and c the intercept. Non linearity is often due to the presence of non-linear elements, *e.g.* electronic amplifier, mechanical hysteresis. Creep and friction of the mechanical parts also contribute to non-linearity, which is expressed as a percentage of deviation from the linear value or from the closest fitting theoretical straight line

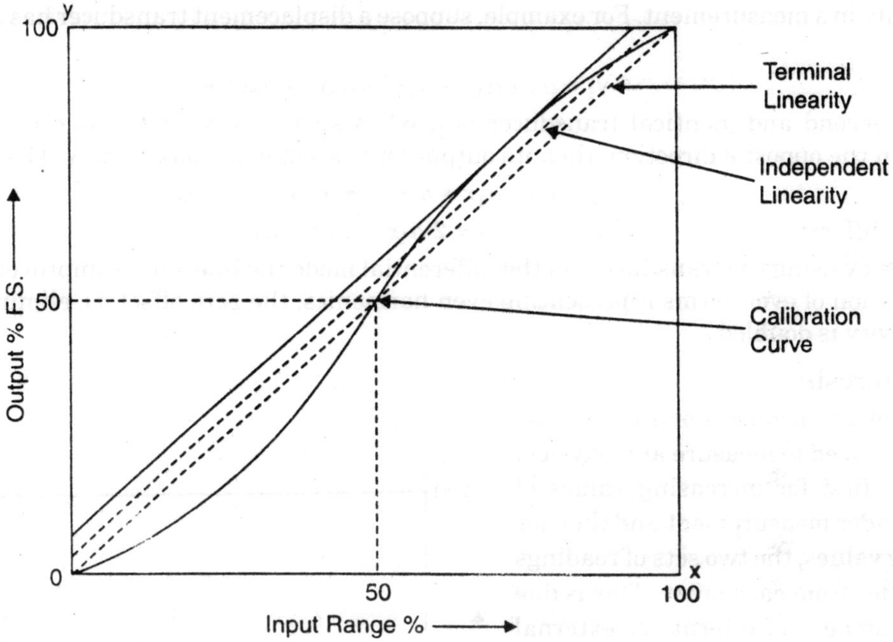

Fig. 2.2. (*a*) Terminal and independent linearity.

relationship. In order to clearly define the term "linearity" three different versions of linearity are usually given, as shown in Fig 2.2,

(*a*) Depicts what could be called "Terminal linearity" *i.e.* the straight line obtained by joining the end points 0% and 100% of the full scale output.

(*b*) Indicates the "Independent linearity", which is the best straight line drawn when all the output readings as well as the end points are covered and

(*c*) "Least square linearity" refers to the straight line for which the sum of squares of residuals are minimised. The residuals are deviations of the output readings from their corresponding points on the best fit straight line. A

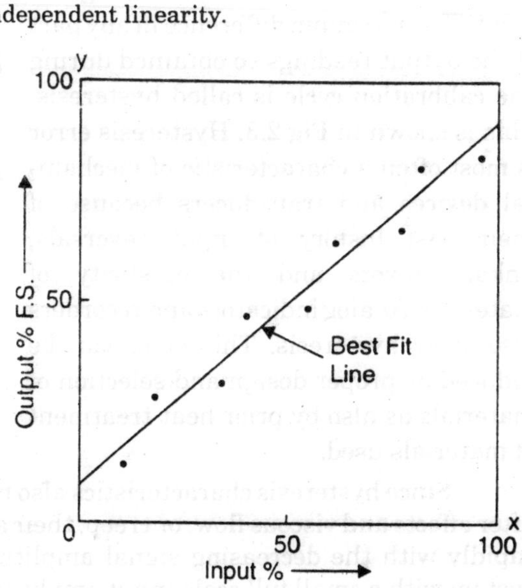

Fig. 2.2. (*b*) Least square fit linearity.

relative term is "scatter" giving the deviation of the mean value of the repeated measurements from the best fit line.

The presence of non-linearity can be particularly serious if it is desired to carry out Fourier analysis of a periodic waveform. The harmonics generated by the lower frequency components can seriously affect the measurements at the higher frequencies, where it is impossible to distinguish between harmonics generated by non-linear effects and the true harmonics of the signal waveform.

Making use of a differential measurement system can sometimes significantly reduce the non-linearity in a measurement. For example, suppose a displacement transducer has an output given by

$$y_1 = a_0 + a_1x + a_2x^2 + a_3x^3 + a_4x^4 + ...$$

If a second and identical transducer is used in such a way that it is connected and displaced in the opposite direction, then its output for the same displacement will be

$$y_2 = a_0 - a_1x + a_2x^2 - a_3x^3 + a_4x^4 - a_5x^5 \text{ ......}$$

The difference $\Delta y = y_1 - y_2 = 2(a_1x + a_3x^3 + a_5x^5 + ...)$

Thus by using the transducers in the differential mode the linearity is improved through the elimination of even terms representing even harmonics, the zero offset a_0, eliminated and the sensitivity is doubled.

2.2.5. Hysteresis

When a-transducers or any measuring device is used to measure any physical parameter, first for increasing values of quantity under measurement and then for decreasing values, the two sets of readings usually differ from each other. This is due to certain amount of internal or external friction in the response of the sensing element. The maximum difference in any part of the output readings so obtained during one calibration cycle is called hysteresis. This is shown in Fig 2.3. Hysteresis error is most often a characteristic of mechanical devices and transducers because of their past history of input reversals, linkage pivots and the elasticity of materials. Analog indicators and recorders also show hysteresis. This error can be reduced by proper design and selection of materials as also by prior heat treatment of materials used.

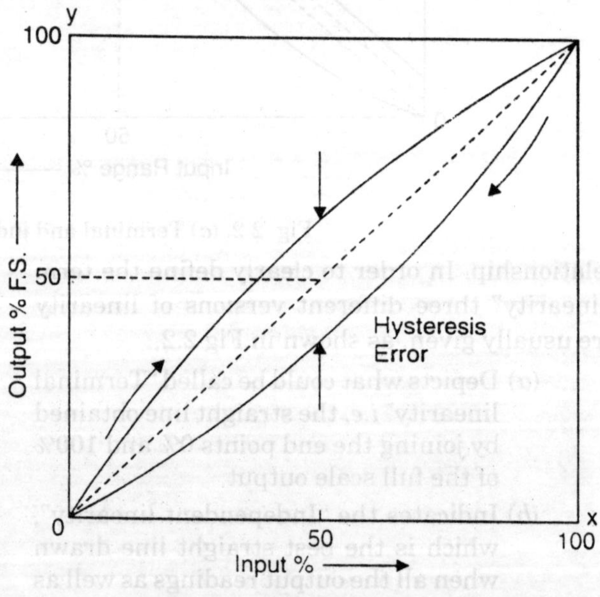

Fig. 2.3. Hysteresis error.

Since hysteresis characteristics also result from backlash in mechanical coupling, elastic after effects and viscous flow, or creep, their absence causes hysteresis losses generally decrease rapidly with the decreasing signal amplitude. Hence it would seem desirable to design the system with a small full scale input amplitude. On the other hand, both noise and backlash are normally independent of the signal level, so that their effect on the output signal becomes more significant as the full scale amplitude is decreased. Evidently, a compromise is necessary to minimise the total distortion. However, even with the most carefully designed system, it is impossible to avoid transient input signals that grossly overload the system. In presence of hysteresis, such signals can permanently affect the static calibration.

2.2.6. Amplitude Distortion

In general, the frequency bandwidth of the measuring system should be somewhat greater than that of the signal. The amplitude response should normally be flat and reasonably free of resonance peaks, preferably with a damping factor in the range 0.6 – 0.8. This gives a reasonable compromise between a fast rise time and freedom from amplitude distortion. For measurements in which significant noise originates either from the signal source or the transducer signal conditioner system, too wide a bandwidth can be a disadvantage in that the extra bandwidth contributes little to the signal but lets extra noise also pass through, thereby increasing the signal to noise ratio. It would thus appear reasonably desirable to design the system bandwidth so that it carefully matches with that of the signal. However in the region in which the amplitude response of the system decreases, phase angle between the input and output also changes rapidly. Such a phase shift can by itself introduce distortion of the signal waveform. In order to avoid this, system bandwidth should be kept much greater than the signal. In case this is not possible due to noise considerations, careful consideration must also be given to phase distortion.

2.2.7. Phase Distortion

Phase angle between input and output depends on frequency. So long as phase angle *vs* frequency characteristic is linear, no distortion of input signal takes place. It simply causes

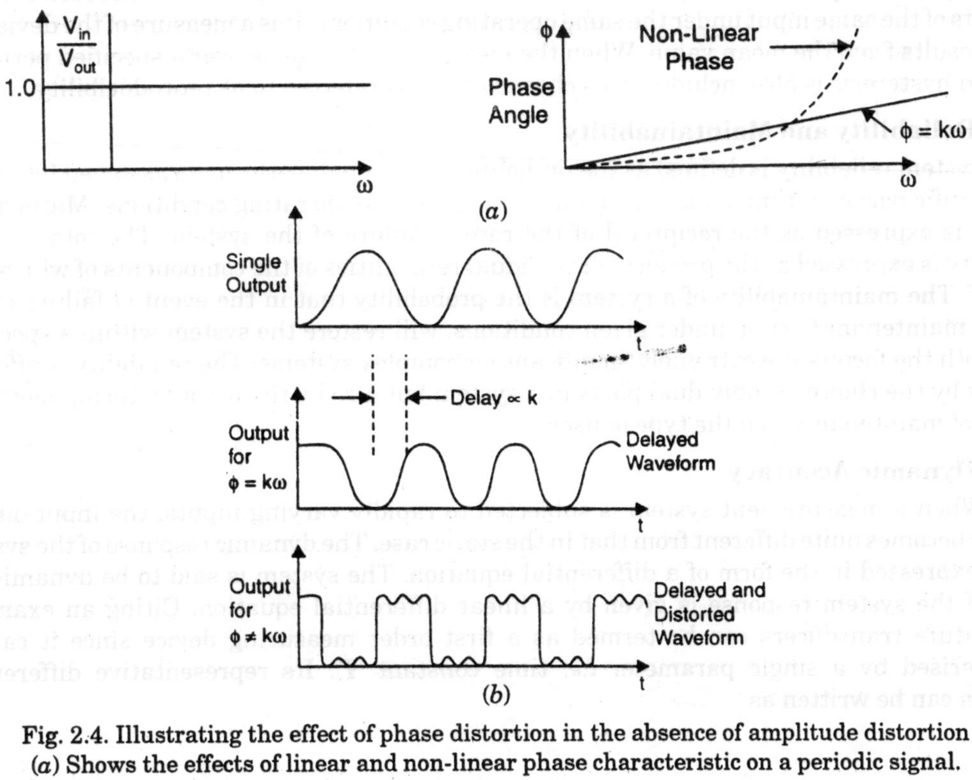

Fig. 2.4. Illustrating the effect of phase distortion in the absence of amplitude distortion (*a*) Shows the effects of linear and non-linear phase characteristic on a periodic signal.

displacement of output in time. As shown in Fig 2.4, the output will then be identical to that with zero phase angle but delayed in time by an amount that depends on the slope of the phase charateristic. A non linear phase characteristic results in the output signal spectral components having the wrong time relation ship, thereby distorting the output signal Fig 2.4 (*b*).

2.2.8. Resolution

Resolution denotes the ability of a measuring device to distinguish between the two equal values. It could be defined as the smallest variation in output value that could be detected when the input is being varied unidirectionally, or the input increment that produces a definite perceivable change in output. The readability of scale is a characteristic that depends on the instrument or the measuring device and the observer. It indicates the number of significant figures that can be read or recorded. In case of analog indicators, readability factor depends upon the scale markings and the thickness of the deflection pointer. In digital indicator, the least significant digit gives the measure of scale readability.

2.2.9. Threshold

It is very often observed that by gradually increasing the input to an instrument, transducers or any measuring device, there is certain minimum value below which no change in output is observed. In other words, the first detectable output change is observed only at a certain minimum value. This minimum value is called threshold. Numerically, it is the minimum output value for a corresponding change in input. Many a times, this threshold value is termed as "Dead band" or "Dead zone". These form a total range of input values possible for a given output and contribute to the total hysteresis.

2.2.10. Repeatability

Repeatability denotes the closeness of agreement among a number of consecutive measurements of the same input under the same operating conditions. It is a measure of the deviation of test results from the mean value. When the measurement is made over a specified period of time and hysteresis is also included, this characteristic is referred to as reproducibility.

2.2.11. Reliability and Maintainability

System reliability is defined as the probability that it will perform its assigned functions for a specific period of time under given environmental and operating conditions. Mathematically, it is expressed as the reciprocal of the rate of failure of the system. The total system reliability is expressed as the product of individual reliabilities of the components of which it is made of. The maintainability of a system is the probability that in the event of failure of the system, maintenance action under given conditions, will restore the system within a specified time. Both the factors are extremely significant for complex systems. The reliability is affected not only by the choice of individual parts in a system but also by the manufacturing methods, quality of maintenance and the type of user.

2.2.12. Dynamic Accuracy

When a measurement system is subjected to rapidly varying inputs, the input-output relation becomes quite different from that in the static case. The dynamic response of the system can be expressed in the form of a differential equation. The system is said to be dynamically linear if the system response is given by a linear differential equation. Citing an example, temperature transducers can be termed as a first order measuring device since it can be characterised by a single parameter *i.e.* time constant '*T*'. Its representative differential equation can be written as

$$y + T\, dy/dx = x(t)$$

where x is the input, y the output and $x(t)$, the time function of input.

Coming to the second order transducers, for example, the spring-mass system employed for acceleration and force measurements, the two parameters that characterise it are the natural frequency f_n (or angular frequency $w_n = 2\pi f_n$), and the damping ratio 'z'. With these two parameters the second order differential equation can be written as :

$$(1/w_n^2)\, d^2y/dx_2 + (2z/w_n)\, dy/dx + y = x(t)$$

where w_n is in radians per second and z is a non-dimensional quantity.

Higher order systems result when more than one lower order systems are joined in series. For example, if the output of a second order transducers is fed to a second order filter, the two together constitute a fourth order system.

The first order and second order system parameters are useful for analysing the output response to simple input-time functions and for evaluating the dynamic errors arising out of them. The lower time constant parameter in case of a temperature sensor indicates a fast response and thus minimum dynamic errors. In the second order system, the natural frequency is an index of speed of response. The damping ratio is a measure of system stability. An under damped system displays oscillatory output in response to a transient input, whereas an overdamped system exhibits sluggish response, thereby taking considerable time to reach the steady rate value.

2.2.13. Sampling Error

The process of digitising an analog signal involves sampling the signal at a fixed frequency and converting the sampled amplitude into the required digital format. For converting the digital signal back to analog requires the reconstruction of the sample amplitude and the use of filter for interpolating between successive sampled amplitudes. If the error involved in converting the sampled amplitude into digital code and back again into an amplitude can be neglected, then the only sources of errors are those arising from the sampling process and from the means of reconstructing the analog signal from the sampled amplitudes. Under these circumstances, it is important to know what sampling frequency should be chosen so that the RMS error between the original signal and that reconstructed will not exceed a given value.

Although the sampling frequency to be used as per the sampling theorem gives a minimum of two f samples per second for a frequency f (under ideal conditions), it is of little practical use in case of a signal containing not single but a spectrum of frequencies. Also, if a signal has spectral components beyond those of practical interest, aliasing or foldover errors can occur which introduce errors in the reconstructed waveform at the frequencies of interest. Such errors can be avoided either by inserting a filter into the signal path to cut out the higher unwanted components or by using a higher sampling frequency.

2.2.14. Effects of Input and Output Impedance

The loading affect imposed by the input impedance of a transducer can cause the measured signal to differ from the undisturbed value. If the signal source impedance is known, corrections can generally be applied. However in some cases *e.g.* in various physiological measurements, it is not possible since the signal impedance can only be estimated, that too only roughly. In addition, signal source impedance frequently varies non-linearly over the time course of measurements. For these reasons, it is desired that the transducers system loading effects be small.

Change in the input and output impedances of a measurement system to meet the specifications demanded by the source of the measured quantity or the load into which the output is delivered, can be achieved sometimes through modifications in design. But it cannot be done without compromising other significant parameters. In such cases, use of feedback enables us to obtain very large changes in the input and output impedances.

In electrical measurements, current feedback from the output lowers both the input and output impedances, on the other hand, voltage feedback increases the input impedance but lowers the output impedance. If the feedback is large, the input transducers act simply as a sensitive null detector to detect if a difference between the feedback quantity and the measured quantity exists, and if it does, to alter the feedback quantity in such a way as to improve the null.

Feedback can be advantageous in mechanical measurements. For *e.g.* suppose it is desired to measure a force whose true value can only be obtained when displacement to be produced by it is negligibly small. Such a measurement can be made by using a force generator to exactly balance the measured force. The difference between the two forces is detected by a displacement transducer and the associated amplifier whose output drives the force generator. The more sensitive the displacement transducer-amplifier system, the smaller will be the null displacement and hence the more accurate will be the measurement.

The feedback from output to input lowers the output impedance. The reason being that when the output of a feedback measurement system is heavily loaded, the amount of signal feedback decreases. This tends to increase the net input signal to compensate for the loading effect. If the compensation is perfect, the output impedance will be zero.

2.3. ELECTRICAL TRANSDUCER

As already mentioned above, the significant parameters that dictate the capability of a transducer are linearity, repeatability and reliability. Based on these criteria, the main advantage of an electrical transducer can be summed up as follows :

(*i*) The electrical output can be amplified easily, to any desired level.

(*ii*) Remote recording and indication of output is possible.

(*iii*) The electrical output can be easily utilised for control purpose in any modified form.

(*iv*) The analog signal can be converted into pulse or frequency form. The analog to digital conversion is also possible for print out, display and online computation.

(*v*) Since the output signal can be modified, modulated or amplified suitably, it can easily be adapted for recording on a multi channel recording oscilliographs, thus catering to a number of electrical transducers simultaneously.

(*vi*) Any combination of outputs from similar transducers or control signals can be obtained by mixing the signal.

(*vii*) The size and shape of transducers can be suitably designed to have an optimum weight and volume. Size can even be made extremely small, thereby increasing the natural frequency to a high value. An example being the piezo electric pickup employed for vibration measurements.

The best accuracy achievable with electrical transducers is of the order of 0.1%. With the availability of better materials, improved technology and circuitry, the range of accuracy and stability have been very much enhanced.

In null or zero indicating systems, feedback circuitry can be employed to get better accuracy, but this also increases circuit complexity alongwith more space, lower natural frequency and higher cost.

2.4. CLASSIFICATION

2.4.1. Active and Passive

As mentioned earlier in this chapter, the transducers are generally of two main types : Active or self generating type which operate under "energy conversion" principle and generate an equivalent electrical signal output, without any external excitation. Passive transducers are those which require external excitation and operate under "energy controlling" principle. These depend upon change in the electrical parameter like resistance, inductance or capacitance, which call for an external source of excitation.

The following table shows a broad classification of transducers:

Active transducers	*Passive transducers*
Piezoelectric Thermoelectric Photovoltaic Magentoelectric Electrodynamic Electromagnetic Pyroelectric	Resistive Inductive Capacitive Piezo resistive Magneto resistive Thermo resistive Photoconductive Elasto resistive Hall effect Synchro Ionic conduction

2.4.2. Primary and Secondary

Just as active and passive transducer was a classification from a functional standpoint, primary and secondary is a classification from an operational stand point. The primary and secondary transducting role can be understood from an example : In a diaphragm type pressure transducer, the pressure displacement of the diaphragm is the primary transducer, while the secondary transducer is the device designed to measure the diaphragm displacement under pressure and to give an output that is usually in the form of electrical signal. That way, many other mechanical-electrical transducers are of this form.

2.4.3. Digital and Analog

An additional classification is sometime useful in describing a transducer from the form of output signal it gives. Although most transducers give an output in an analog form, some of them are designed specifically to yield an output that is either in a digital form or can be encoded into a digital form. It must be accepted that most of the digital transducers are essentially comprised of an analog-analog transducer with a built-in analog to digital converter. However, for a transducer to be described as digital, its output must either be in a digital form or in a form that is readily converted into it. For example, a temperature transducer based on the change in resonance frequency of a quartz crystal with temperature is usually classified as a digital transducer, since the output is in the form of frequency of oscillation, which can be readily measured with a digital counter. A still clearer case is that of a displacement transducer using a encoded plate with transparent and opaque regions. If such a plate is moved past a set of optical detectors, a digital output in the form of pulses proportional to displacement can be obtained. This type of transducer is most frequently used for shaft angle (rotation) measurements.

2.5. SELECTION CRITERIA FOR A TRANSDUCER

A transducer normally designed to respond to only one specific measuring parameter. Hence total information about its electrical and mechanical characteristics is of great significance while selecting one for a particular application. While choosing a transducer, the following are to be kept in mind :

(*i*) Nature of the physical quantity to be measured.
(*ii*) Choice of a best suited transduction principle.
(*iii*) Compatibilty of input-output characteristics with indicating and recording systems.
(*iv*) Ruggedness to withstand overloads and safety measures against overload protection.
(*v*) Linearity : Output varying linearly and symmetrically with input.
(*vi*) Repeatability : Ability to reproduce output signals exactly when the measurand is applied repeatedly under the same environmental conditions.
(*vii*) High reliability and stabilty : measurement unaffected by environmental conditions, natural and man made.
(*viii*) Good dynamic response : Output faithful to input when taken as a function of time.
(*ix*) Excellent mechanical characteristics, deviod of :

- **Mechanical hysteresis :** a measure of imperfect response of sensing elements, integrated over dimensions of strained transducers, also depending upon material and ageing.
- **Creep :** due to viscous flow in the material of sensing elements, effect of creep increases with increasing load and temperature. Material with low melting point show larger creep.

BIBLIOGRAPHY

1. Dr. Harsh Vardhan, *Measurements : Principles And Practices*, Macmillan (India) Ltd. 1993.
2. C.S. Rangan, G. R. Sarma and V.S. Mani, *Instrumentation Devices and Systems*, Tata Mcgraw Hill 1983.
3. D.M. Considine, *Process Instruments and Controls Handbook*, Mcgraw Hill, New York 1974.
4. N.N. Norton, *Handbook of Transducers For Electrical Measuring Systems*, Prentice Hall, New Jersey 1968.
5. Richard S.C. Cobbold, *Transducers for Biomedical Measurements : Principles and Applications*, John Wiley And Sons 1974.
6. Schular A.E., *Digital transducers (Paper 662) in Advances in Instrumentation*, Vol xxiv pt. 4. Proc. 24th Annual ISA Conference, 1969, Instrument Society of America, Pittsburg, 1969, p. 1-8.
7. G.F. Harvey, Ed. ISA, *Transducers Compendium*, 2nd Ed. Pts. 1-3 Instrument Society of America : Pittsburg, 1969-1972.

3

Resistive Transducers

This chapter is devoted to a class of transducers based on the variation of electrical resistance with some change in a physical quantity. Covered under these are resistive transducers for measurement of displacement, both linear and angular. Strain gauges are also passive transducers wherein a resistance element whether wire type, foil type or semiconductor type is allowed to undergo a strain or a temporary physical change within elastic limits of the member to which these are attached. In the process, a change in resistance or material resistivity may occur which gives a measure of displacement, stress (force per unit area), force or pressure. Such basic measurements are of interest in civil, mechanical or structural engineering applications in order to determine the normal and safe load bearing capacity of the mechanical components of the structures. The area of application of strain gauges is so wide that it requires a separate and exhaustive treatment.

Again, in the temperature measurements, resistance thermometers can be considered in the broad category of resistive transducers because these ultimately depend upon change in resistance of the element when exposed to temperature.

The above categories are now taken one by one in manner described above.

3.1. POTENTIOMETRIC TRANSDUCERS

One of the simplest and most efficient forms of resistive transducers is the variable resistance potentiometer with a movable sliding contact. This is most often used for measuring displacement. As shown in Fig. 3.1, the movement could be translational as well as rotational. The former normally covers a linear displacement range from 0.2 to 150 cm and the latter has an angular displacement range from 10 degrees to 340 – 350 degrees. As is obvious, change in displacement produces a resistance variation. Earlier, wire wound resistance elements were

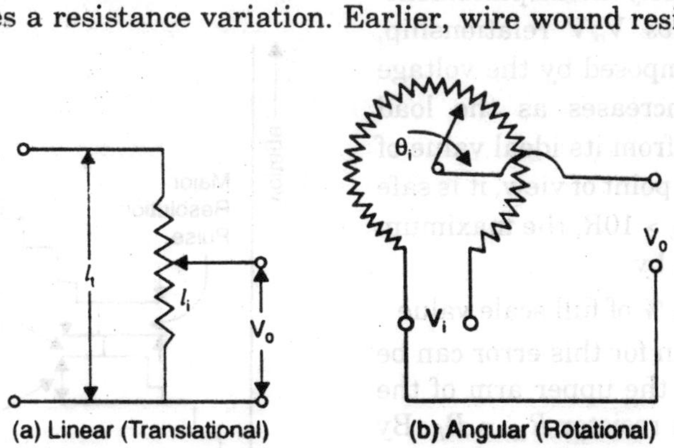

Fig. 3.1. Potentiometric displacement transducer.

normally used but later, these gave way to conducting ceramics, *e.g.* cermet or conductive plastic resistance elements, which give performance better than even the best of wire wound resistors. Referring to Fig 3.1 (*a*), $V_o = (l_i/l_t)V_i$ or $l_i = (V_o/V_i)l_t$ (assuming ideal linear response), and referring to Fig. 3.1 (*b*), $V_o = (\theta_i/\theta_t)V_i$ or $\theta_i = (V_o/V_i)\theta_t$.

In practice, output terminals are connected to an electrical indicating instrument that introduces loading error caused by its input impedance or resistance R_L. Major drawbacks of

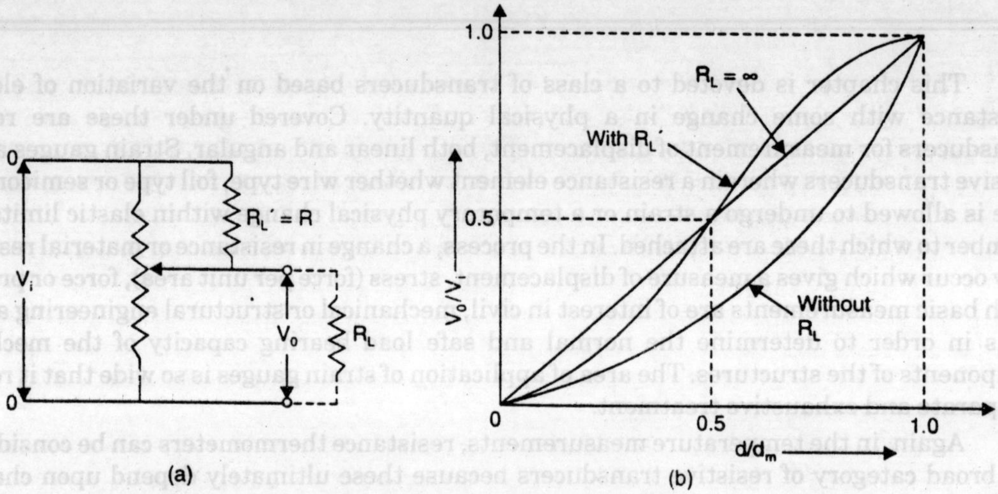

Fig. 3.2. Effect of load resistance on linearity of potentiometric transducer.

potentiometric type resistive transducers are poor dynamic response *i.e.* response under rapidly varying inputs, susceptibility to vibration or shock and poor resolution. So far as the loading error is concerned, the measuring instrument or device should ideally have an infinite resistance.

The non linearity in displacement-resistance or d/d_m *vs* V_o/V relationship, caused by the load imposed by the voltage measuring device increases as the load resistance decreases from its ideal value of infinity. From design point of view, it is safe to assume that for $R_L > 10R$, the maximum loading error is given by

$$\text{Error}_{\max} = 15R/RL\ \%\ \text{of full scale value}$$

A compensation for this error can be effected by shunting the upper arm of the potentiometer with a resistor $R_L' = R_L$. By doing so, the loading error can be reduced to zero at a displacement value $d = 0.5$ dm. Another consideration in the design of potentiometric type displacement

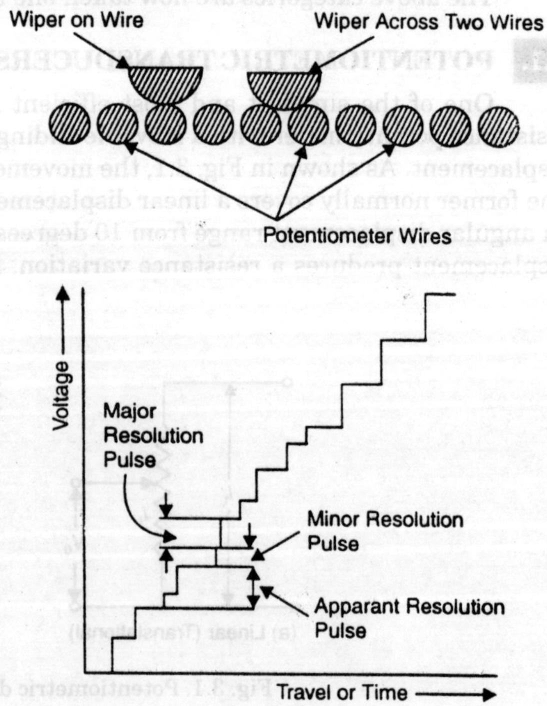

Fig. 3.3. Effect of finite wire diameter on the resolution of a wirewound potentiometer transducer.

transducer is its sensitivity or resolution. As shown in Fig 3.3, when the wiper or the variable contact moves across the wire, it produces a small step change in resistance. A practical limit in wire wound potentiometers is usually 500 turns/cm. This limits the ultimate value of resolution achievable with a translational type potentiometer to 20 μm. This effect is noticeable in wire wound potentiometers only.

3.1.1. Resistive Strain Gauges

The potentiometric devices measure displacement without altering the physical properties of the resistance element itself. Owing to various limitations of such devices (mentioned above), these are not very much used. In this section, we examine such transducers, which undergo physical change in resistance arising as a result of strain produced by applying displacement, force or pressure. These are called strain gauges. A fractional change in length produced by the application of a stress or the force per unit area, is termed strain, *i.e.* strain $= \Delta L / L$ where ΔL denotes change in length L. Strain is therefore a ratio having no units. Since resistance is directly proportional to length, ΔL will produce a corresponding change in resistance ΔR. Obviously, strain is proportional to stress within elastic limits hence, change in resistance can be taken as a measure of stress, or force per unit area.

Depending on this construction, strain gauges can be produced in either bonded or unbonded forms. The unbonded gauge consists of a wire 0.002 cm or less in diameter wound between a fixed and movable frame. As shown in Fig. 3.4, there are four sets of identical wires wound under tension so that the movable frame can be displaced by a small amount in either direction without reducing tension in any of the wires to zero. To measure change in resistance, the four strain gauges are connected in a whetstone bridge as indicated. A motion to the left causes the resistance R_1 and R_4 to increase and R_2 and R_3 to decrease thereby yielding a bridge output voltage twice that of a two wire system. A further advantage of this arrangement is self-compensation for environmental effects. Changes in temperature affect each wire to almost the same extent, thereby leaving the output voltage unchanged.

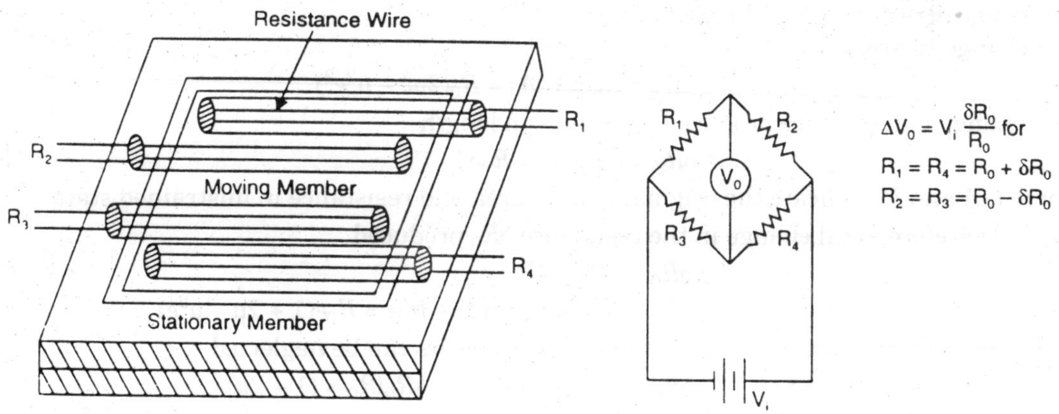

Fig. 3.4. Temperature compensated unbonded strain gauge and Wheatstone bridge.

The unbonded strain gauge is very efficient in that force applied to its ends is not shunted by any mechanical members. In spite of its apparent fragility, it is widely used as a secondary transducer for force, pressure and acceleration measurement. Many earlier types of diaphragm operated blood pressure monitors used this transducer element.

A bonded strain gauge, on the other hand, consists of a metal wire, etched metal foil, vacuum deposited film of metal or a semiconductor bar attached to a suitable backing material.

They are designed to be cemented to the surface whose strain is under measurement. Once cemented, these effectively become part of the surface, so that ideally, any strain that occurs in the surface is transmitted unchanged to the gauge element. Fig. 3.5 shows some arrangements of bonded wire and foil strain gauges.

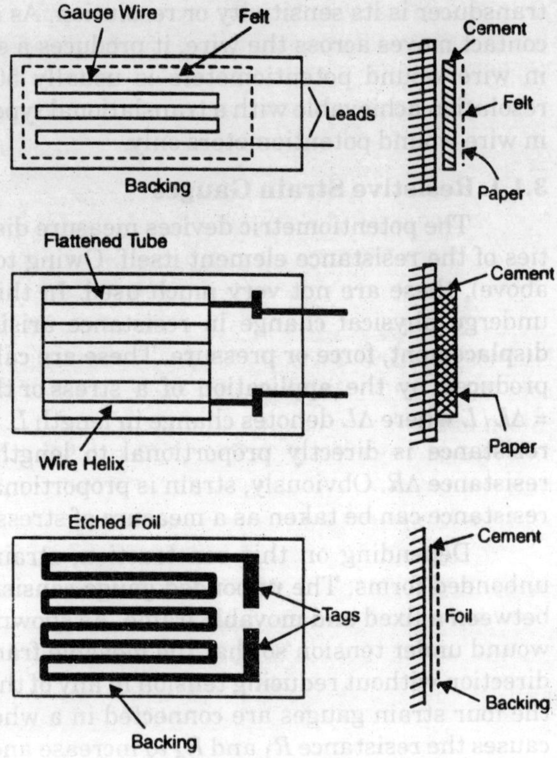

Fig. 3.5. Bonded strain gauges
(*a*) Flat grid type, (*b*) Wrap around type, (*c*) Etched foil type.

In practice, when a load is applied, the strain gauge gets strained along with the surface to which it is bonded. The resistance changes on account of change in length as well as diameter. When a strain is produced lengthwise, a strain of lesser magnitude but with opposite sign is also produced in a perpendicular plane, the ratio being μ, the Poisson's ratio. μ is a function of the material of which the gauge is made of ; 0.3 for most materials.

Let e be the strain in length $\delta l/l$ then the change in cross sectional areas,

$$\delta A = (\pi/4)d^2 - (\pi/4)(d - \mu e)^2$$

where d is wire diameter

or $\delta A = (\pi/4)d^2\,(2\mu e - \mu^2 e^2)$

Since the resistance increases with l increasing and also with A decreasing, the change in resistance δR_1 brought about by a change in area,

$$\delta R_1 = R_0(\delta A/A) = R_0(2\mu e - \mu^2 e^2)$$

and, change in resistance produced by change in length

$$\delta R_2 = R_0(\delta l/l) = R_0 e$$

where l, A and R_0 indicate the original length, area and resistance in unstrained state.

Therefore, total change in the resistance δR_0 produced

$$\delta R_0 = \delta R_1 + \delta R_2$$

$$= R_0(2\mu e - \mu^2 e^2) + R_0 e = R_0 e(1 + 2\mu - \mu^2 e)$$

Since e and μ are very small, their higher powers can be neglected

Therefore $\qquad \delta R_0 = R_0(1 + 2\mu)e$ or $\delta R_0/R_0 = (1 + 2\mu)e$

In the above derivation, we have assumed that resistivity ρ which is a property of material (irrespective of the wire dimensions), remains constant and unchanged. However in reality, the lattice structure of the material also gets altered by strain, the resistivity ρ of the material may also change and this in general, causes the resistance to increase further. Therefore the above relation will also carry an additional term or

$$\frac{\delta R_0}{R_0} = (1 + 2\mu)e + \frac{\delta\rho}{\rho}$$

the first term on the right hand side represents the dimensional effect and the second term, piezoresistive effect.

The gauge factor or the strain sensitivity of a gauge is defined by

$$G = \frac{\delta R_0/R_0}{e} = (1 + 2\mu) + \frac{\delta\rho/\rho}{e}$$

Since $\mu = 0.3$ for most of the materials, this reduces to

$$G = 1.6 + \frac{\delta\rho/\rho}{e}$$

The gauge factors for strain gauges made of Constantan (Ni 45%, Cu 55%) and Karma (Ni 74%, Cr 20%, Fe 3%, Cu 3%) are 2.1, while that for Nichrome 5 (Ni 30%, Cr 20%) ranges from 2.1 to 2.63. However, in case of single crystal semi conductor strain gauges made out of Silicon and Germanium, the gauge factors are much higher, in the range of 100 to 170 since the piezoresistive term in these becomes predominant. Though these are 70-75 times more sensitive than the metal strain gauges, they do suffer from much higher resistivity temperature coefficient, making it essential that steps be taken in the design to provide temperature compensation.

Temperature variation is, in fact a major source of error in measurements with a strain gauge. Temperature variation acts in two ways :

1. It alters the gauge resistance.

2. It causes differential expansion between the gauge and the deformable material surface or the test piece, thus causing variation in the measurable strain. In general, it is always not possible to apply correction for the temperature effect. The usual method is to install two identical strain gauges ; one active and another one dummy. Temperature variation would equally affect both, thereby compensating each other, since both are at the same temperature. Fig. 3.6 shows how these are connected in two opposite arms of a wheatstone bridge. The resistors R_1 and R_2 are chosen to have a negligible temperature coefficient.

At balance, $R_1/R_2 = R_{ag}/R_{dg}$

Other sources of error are :

A. Humidity. This causes breakdown of insulation between the gauge and the ground point, as also the electrochemical corrosion of the strain gauge wire due to electrolysis whereby the gauge resistance tends to increase considerably and a zero drift occurs. An appropriate protective coating helps minimising these effects.

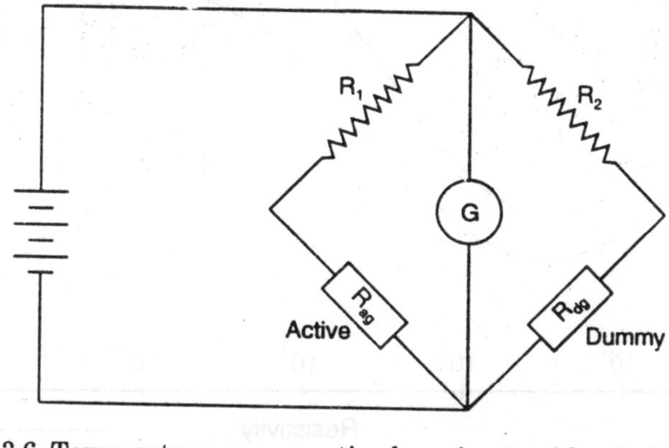

Fig. 3.6. Temperature compensation by using two identical strain gauges connected in a Wheatstone bridge.

B. Magnetostrictive effect. This effect is observed in strain gauges made of ferromagnetic materials. This effect produces disturbing voltages as large as $2mv$. The gauge, if properly conditioned by repeated straining and passing current through it while in strain, may free the gauge from this effect.

3.1.2. Thin Film Or Foil Type Gauges

Lately, evaporated thin film strain gauges are receiving attention. Thin films of Aluminium, Gold, Nickel, Platinum or Palladium are formed by vaccum deposition process, in the desired patterns directly on to a substrate by thermal evaporation in vaccum. This substrate is then attached to the specimen. Gauge factor is defined in the same way but the major contribution in gauge factor G is due to resistivity changes rather than dimensional changes. Specific sheet resistance R_f, which is the basic property of the film, is the contributing factor for the gauge resistance. The film gauge resistance R_g is related to the specific resistance of the film R_f by :

$$R_f = (b/l)R_g$$

where b is the width and l the length of the film.

Since $$R_g = \rho_f l/A, R_f = \rho_f/t$$

Where ρ_f is the film resistivity in ohm-metre, A, the film cross sectional area and t the film thickness. R_f is therefore a measure of film thickness, increasing with decreasing film thickness. The relation is non linear and for very thin films of the order of 50 Å , R_f may be as large as 10^6 ohms. It is therefore possible to obtain thin film strain gauges of high resistance values with smaller dimensions, which gives rise to higher sensitivity and low power consumption. Gauge factor G of thin films of Cobalt and Platinum varies with resistivity as shown in Fig. 3.7. These gauges can operate over a wide temperature range from – 200 C to 400 C, with good stability. Sensitivity is high because of high G and high gauge resistance. These gauges are rugged and have very low hysteresis and creep.

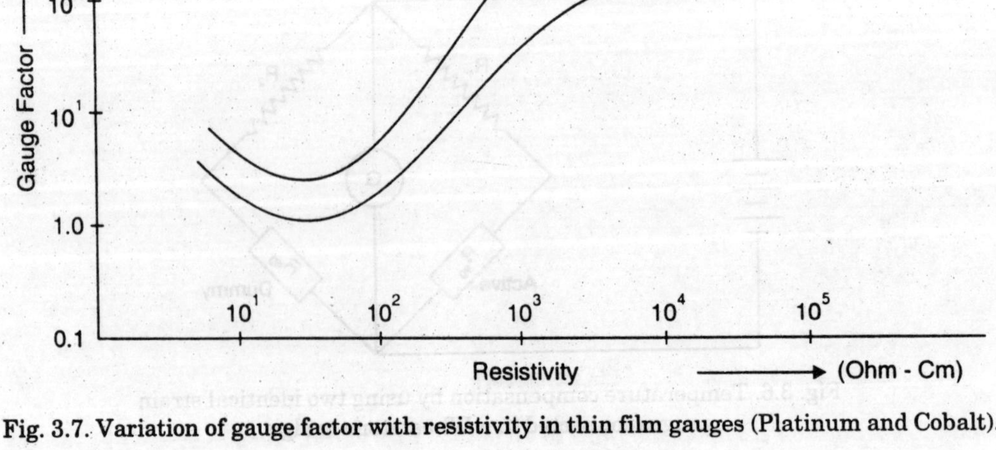

Fig. 3.7. Variation of gauge factor with resistivity in thin film gauges (Platinum and Cobalt).

Hysteresis means that the strain measured should not vary in case of cyclic loading.

Creep. In order to ensure good bonding with faithful strain transmission from specimen to gauge, both backing material and bonding cement/adhesive must be much stronger than filament. If not, the gauge will relax under strain and measured value will be lower, becoming more and more non-linear as strain increases. This is the creeping effect.

Fatigue. In dynamic measurements, fatigue is due to stress reversals occurring at lead junctions. These can cause decay in the base of the gauge or in the binding cement.

Semiconductor strain gauges. The development of Silicon strain gauges around 1960 was a significant advance in the strain gauge technology. Silicon gauges are formed from a single crystal whose orientation and doping are the most important design parameters. The bonded type Silicon strain gauges are made from wafers about 0.2 mm thick with lengths and resistance values nearly equal to the wire type gauge. It uses either Silicon or Germanium base materials to be made available in both *n* and *p* types, *p* type has a positive gauge factor and *n* type has a negative one. These have very large gauge factors varying from – 100 to + 150, but are very brittle and have a high temperature coefficient, often exceeding 0.15/°C. Pure Silicon or Germanium have very high resistivities but when doped with atoms of Phosphorous or Boron, the resistivity falls with increased degree of doping. Usually a large gauge factor is obtained by using lightly doped materials. A high degree of doping also reduces the temperature dependence of the gauge.

Semiconductor strain gauges are inherently more non linear than metal gauges and non linearity increases as doping decreases. The higher non-linearity of *n*-type Silicon or Germanium make it desirable to prefer *p*-type where linearity is of importance. However, the temperature compensation property of *n*-type gauges (having a negative gauge factor), makes it particularly useful when large temperature variations are encountered. The diffused semiconductor gauges are manufactured by using diffusion process as in the manufacture of *IC* chips, by simple cutting or etching a suitably doped single crystal slice in the appropriate direction. Typically, such a gauge may be 0.01 mm thick, 3 mm long and 0.1 mm wide, with a resistance in the range 100-1000 ohms. Alternatively, the gauge may be fabricated by selective diffusion of impurities into a silicon substrate using a SiO_2 mask formed by photolithographic process. The impurities added convert a portion of the substrate into a region of opposite conductivity type, as shown in Fig. 3.8 (*b*). With this method of fabrication, gauges with resistance values greater than 5000 ohms can be made. Much higher impurity concentrations, while retaining a higher resistance value, can be achieved through the process of masked ion implantation. Such a method appears

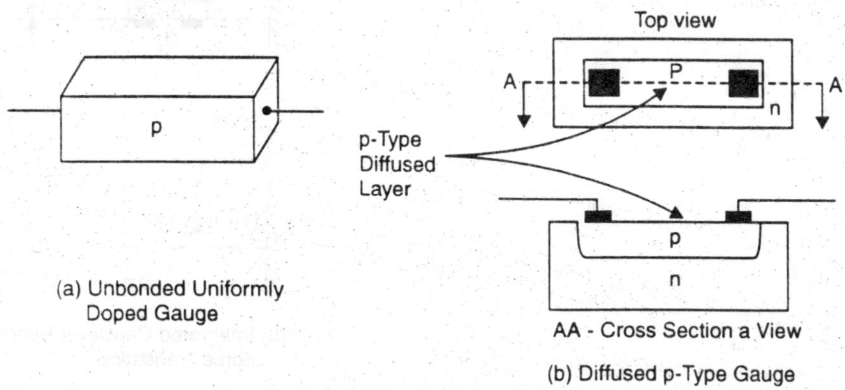

Fig. 3.8. Semiconductor strain gauges.

to have considerable advantages in fabricating Silicon gauges having a nearly temperature invariant gauge factor and resistance.

Both the uniformly doped bar type gauge and the diffused gauge are commercially available with backing material to facilitate attachment to a surface. Alternatively, an unbonded gauge can be attached directly to a surface by using epoxy or nitrocellulose cement or phenolic resin used for bonding.

A Silicon strain gauge has an advantage in that the method of fabrication enables an integrated type transducer to be fabricated in which the Silicon substrate forms the structural member and acts as a primary transducer. The member whose strain is required to be measured *e.g.* a diaphragm in pressure transducer, is formed with Silicon and not a metal and impurities are then deposited into it by diffusion process.

A pressure differential causes the Silicon to bend, setting up a radial stress component at the edge that is opposite in sign to the tangential stress component near the center. Using an arrangement with eight diffused strain gauges, as shown in Fig. 3.9 (*a*), a high sensitivity coupled with good temperature coefficient, can be achieved. Another variation could be a force transducer shown in Fig. 3.9 (*b*), in which the gauges are diffused into the Silicon cantilever

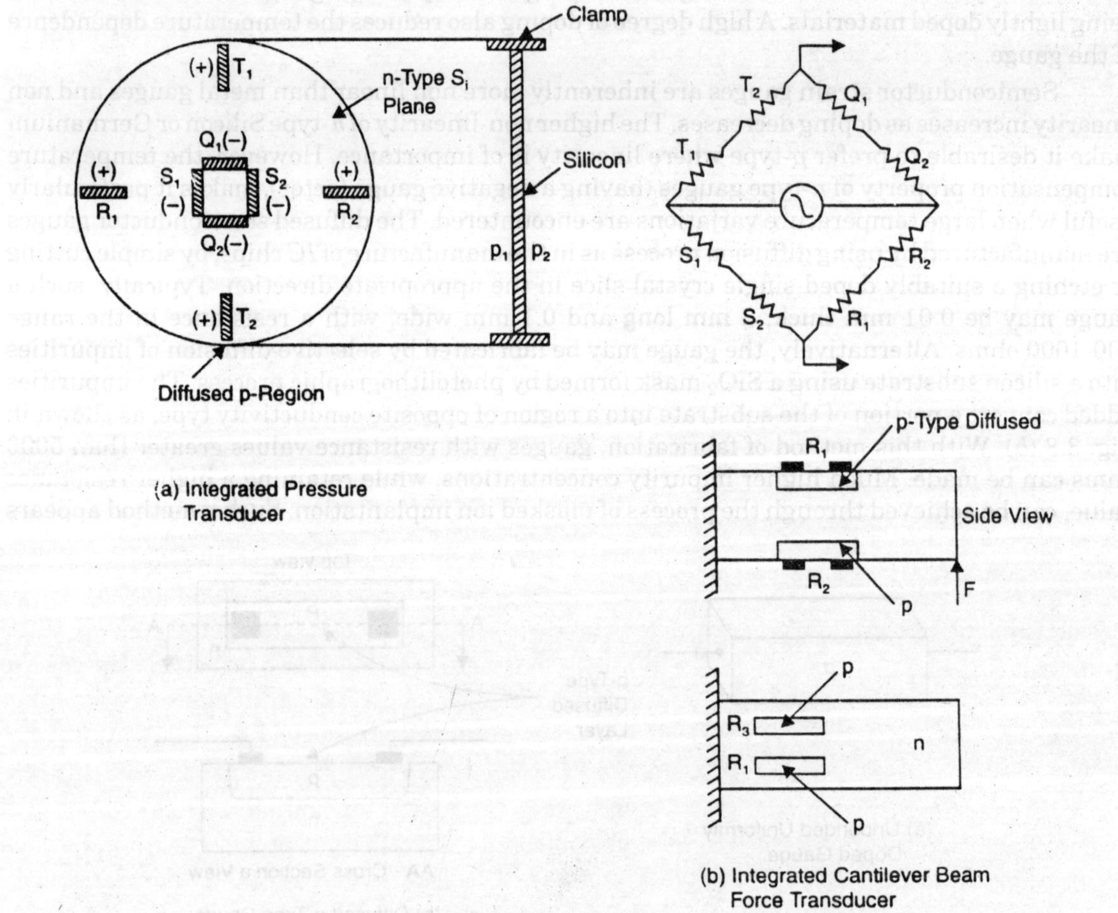

Fig. 3.9. Integrated transducers using semiconductor strain gauges.

beam. Using four gauges, two in compression and two in tension, a linear temperature compensated output is obtained that is proportional to the applied force.

Multidirectional strains. When an elastic member is stressed, and a measure of strain at a single particular point is desired, a strain gauge will measure over a small specified area only. A general measurement problem will arise when the measurement of strain in more than one directions is desired. If a number of gauges could be oriented on an elastic member in such a way that the strains along the principal stress axes could be ascertained, the problem will suitably be tackled. The arrangement of strain gauges for such measurement is known as "Rosettes". These are generally of two types :

- Rectangular rosettes
- Delta rosettes

Fig 3.10 below illustrates the arrangements.

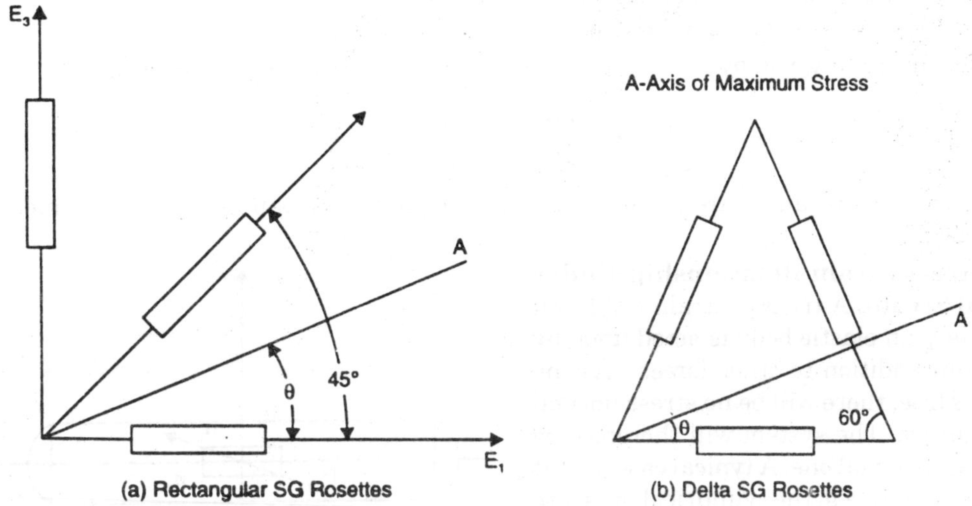

Fig. 3.10. Strain gauge rosettes.

When a gauge is arranged in a direction perpendicular to the direction of principal stress, the strain gauge exhibits what is known as cross sensitivity factor defined as

$$\gamma = \text{Transverse sensitivity/Longitudinal sensitivity}$$

The value of this factor is generally less than 2%. In structural analysis, the direction of principal stress axes may not always be known, hence a complete stress-strain analysis will be possible only with the provision of at least three gauges. The rosettes incorporating two, three or more grids assume importance due to this. Where the direction of two principal stresses are known, a rectangular rosette with a minimum of two elements, at right angles to each other will be called for. However for completely unknown, biaxial stress conditions, a delta rosette will be needed.

Stress-Strain gauges. Stress-Strain gauge is a useful innovation. This is composed of two uniaxial strain gauge arranged at right angles to each other but provided with a common connection terminal. See Fig. 3.11 below :

Connected in this manner, these gauges can be operated either independently or in combined formation. When in combination, these provide an output that is proportional to the stress along the principal gauge axis *a-a*. This is made possible if the ratio of resistances of G_2 to G_1 is equal to the Poisson ratio to which the gauge combination is bonded. The strain along *a-a* or *b-b* is measured in the usual manner with individual gauge factor being considered ,whereas the stress along any axis is measured by using the stress-gauge factor λ_{SG}, usually supplied by the manufacturer. If R_1, R_2 be the resistances of the gauges G_1 and G_2 respectively, ΔR_1, ΔR_2. The change in resistances, when under strain, E, the modulus of elasticity, then the stress along the axis *a-a* is given by :

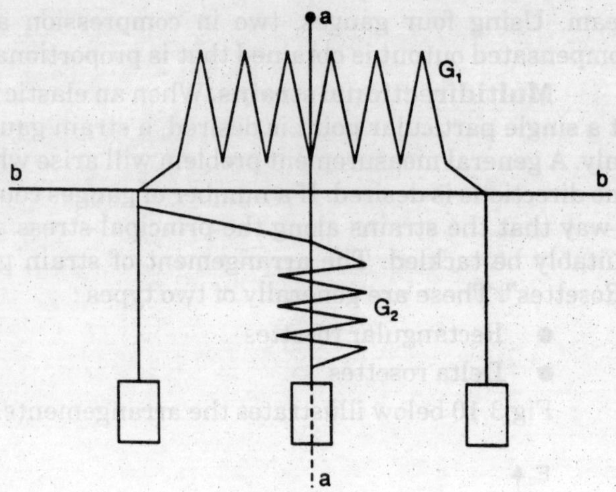

Fig. 3.11. Stress-Strain gauge.

$$S_{a\text{-}a} = \frac{\Delta R_1}{R_1}\left[1 + \frac{\Delta R_2}{\Delta R_1} \Big/ 1 + \frac{R_2}{R_1}\right] E \cdot \frac{1}{\lambda S_G}$$

Stress-Strain gauges are usually of foil type, using constantan with proper compensation for temperature.

Stress-Strain Relationship Under Triaxial Strain. A triaxial strain condition may arise if an elastic body is acted upon by a couple in addition to other forces . As the surface is free, there will be no stress normal to the surface. The system will therefore be treated as a biaxial one. A typical case is that of a regular body, *e.g.* the cylindrical pressure vessel shown at Fig. 3.12.

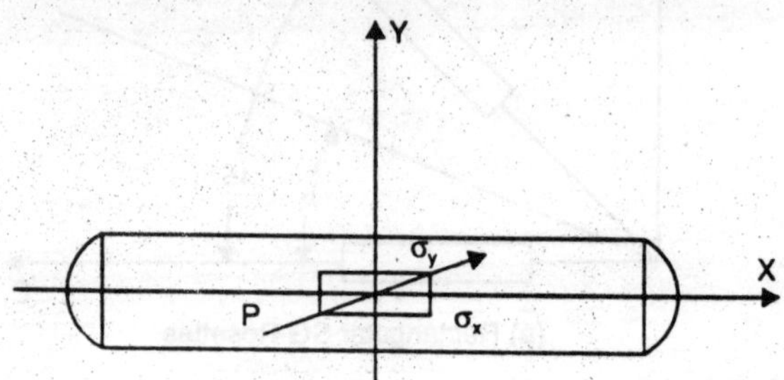

Fig. 3.12. A cylindrical pressure vessel under triaxial stress.

Strain along x and y axes will be given by :

$$\in_x = (\sigma_x - \mu\sigma_y)/E \text{ and } \in_y = (\sigma_y - \mu\sigma_x)/E$$

where E = Young's Modulus

μ = Gauge factor of the strain gauge S_G

σ_x = Longitudinal stress

σ_y = Hoop stress along the circumference.

Solving the above equations, we get

$$\sigma_x = \frac{E(\in_x - \mu\in_y)}{1 - \mu^2} \text{ and } \sigma_y = \frac{E(\in_y - \mu\in_x)}{1 - \mu^2}$$

However as stated earlier, in structural analysis, one does not always know these principal directions of stress. Hence, for a complete stress-strain analysis, a minimum of three strain gauges are necessary instead of a single one.

Strain Gauge Pressure Transducers ; Some Practical Examples

In a typical configuration, a simply supported cantilever acts as an elastic member, to either side of which two strain gauges are fixed so that one acts on tension and the other on compression. A differential pressure actuated diaphragm is attached to the free end of the cantilever by a flexible mechanical linkage. As shown in Fig. 3.13, the differential pressure $\Delta P = P_1 - P_2$ acts as a net force on the area A of diaphragm. The main design requirement is that cantilever stiffness should be higher than that of diaphragm to minimize the effect of hysteresis at low range of stress.

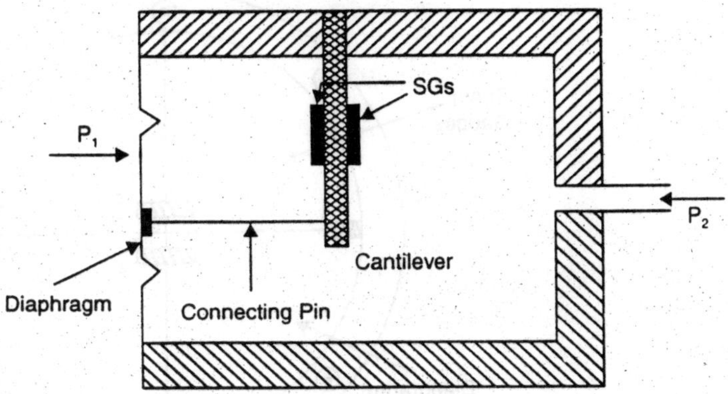

Fig. 3.13. Strain gauge pressure transducer (Diaphragm type).

In an alternate arrangement, four strain gauges could be mounted directly on such points on the diaphragm itself. These points should be so chosen that tensile and compressive stress exist there. For a thin flat diaphragm clamped at the edge, and subjected to a uniform pressure difference ΔP, the tangential stress S_t and radial stress S_r on any point on the low pressure side are given by :

$$S_t = \frac{3}{8}\Delta P\frac{(R^2)}{t^2}\left[(3+\mu)\frac{r^2}{R^2} - (1+\mu)\right]$$

$$S_r = \frac{3}{8}\Delta P\frac{(R^2)}{t^2}\left[(3\mu+1)\frac{r^2}{R^2} - (1+\mu)\right]$$

where μ = Poisson ratio, R = Radius of the diaphragm, t = Thickness of diaphragm
r = Distance of the point from the diaphragm center.

These stresses are valid for small pressures only. As shown in Fig. 3.14, strain gauges 1 and 3 measure radial strain and 2 and 4 measure tangential strain. The corresponding net values of strain are given by :

$$\in_t = \frac{S_t - \mu S_r}{E} \quad \text{and} \quad \in_r = \frac{S_r - \mu S_t}{E}$$

From which,

$$S_t = \frac{E(\in_t + \mu\in_r)}{(1-\mu^2)} \quad \text{and} \quad S_r = \frac{E(\in_r + \mu\in_t)}{(1-\mu^2)}$$

For a clamped circular diaphragm, the maximum radial stress at the edges can be obtained by putting $r = R$

So that

$$(S_r)_{max} = \frac{3}{4}\Delta P\frac{R^2}{t^2}$$

and within the linear range, the strain

$$(\in_r)_{max} = \frac{3}{4}\Delta P\frac{R^2}{Et^2}$$

where E is the Young's modulus of the material.

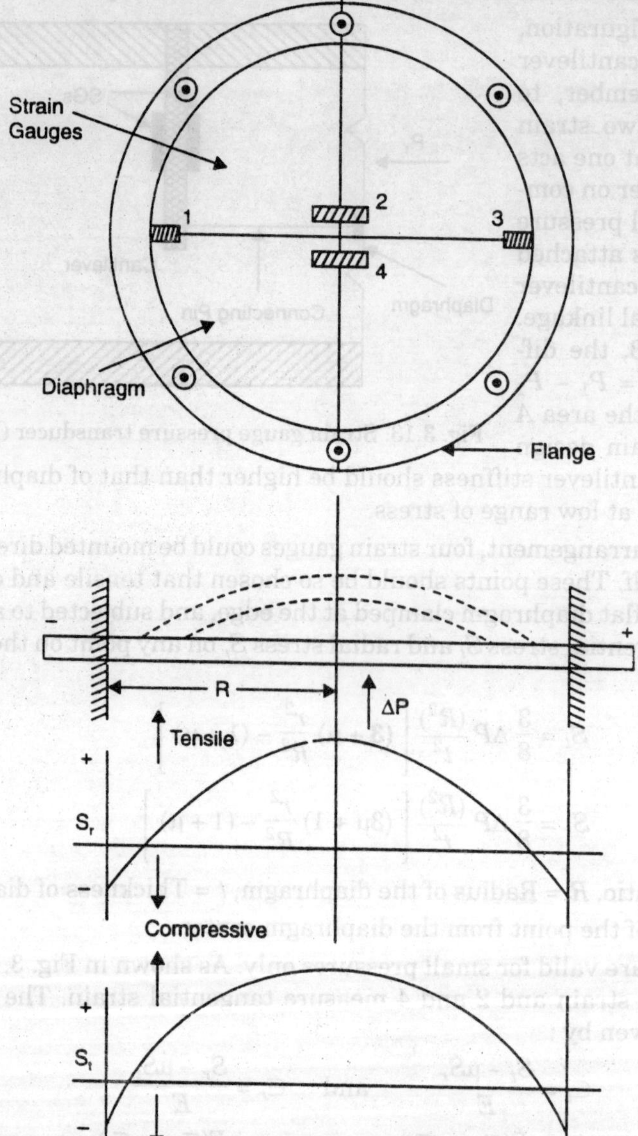

Fig. 3.14. Variation of radial and tangential stresses S_y and S_t with diaphragm radius.

For thick diaphragms, $(R/t \approx 10)$, the calibration would be linear for a pressure upto 70×10^6 pascals but the corresponding stress level at that pressure is quite high imposing a restriction on the higher limit. For a thin diaphragm $(R/t \approx 100)$, the maximum pressure for a linear operation is limited, resulting in low strain values. The diaphragm is mechanically aged for a minimum long term drift.

***IC* technology has provided the scope of integrating semiconductor strain gauges on a silicon diaphragm, along with the additional temperature dependent resistors for temperature**

compensation and operational amplifier for amplification of out-of-balance voltage output resulting from a wheatstone bridge, as shown at Fig. 3.15 below :

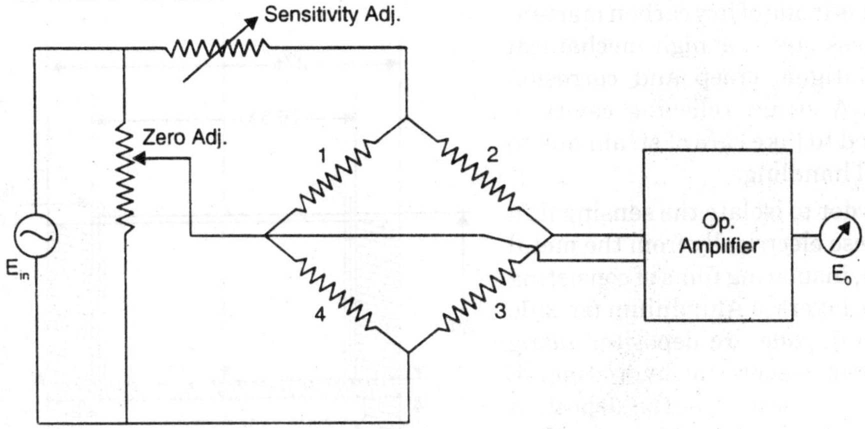

Fig. 3.15. Wheatstone bridge circuit for Fig. 3.14.

The entire system is commercially available in different sizes and when mass produced, their accuracy is ± 2%. Easy repeatability, small size and a large enough amplified voltage output required for data acquisition systems, are the advantages.

3.2. THIN MANGANESE FILM STRAIN GAUGE PRESSURE TRANSDUCERS

During the recent years, a unique development has been reported by using thin manganese film a strain gauge. Manganese film has been employed for sensing strain because it has a very high resistivity and a very low temperature coefficient of resistance compared to all other metals, as shown in the table below. High resistivity results in a high gauge resistance and can thus reduce the errors arising out of resistances of connecting cables, terminals and contacts.

Electrical resistivity and temperature coefficient of resistance (TCR) of different materials

Material	*Resistivity in μ-Ω-cm at 20°C At room temperature*	*TCR (10^{-3} per deg C) in the range 0-100°C*
Aluminium	2.67	4.50
Antimony	40.10	5.10
Bismuth	117.0	4.60
Chromium	13.20	5.90
Copper	1.69	4.30
Gold	2.20	4.00
Manganese	160.0	0.40
Nickel	6.90	6.80
Silicon	2.3×10^{11}	– 81.00
Silver	1.63	4.10
Platinum	10.58	3.92
Cerium	85.40	0.90

A schematic of the diaphragm assembly is shown at Fig. 3.16. The diaphragm is made of low carbon martensitic stainless steel for high mechanical strength, fatigue, creep and corrosion resistance. A strain relieving cavity is incorporated to take care of strain due to mechanical handling.

In order to isolate the sensing film of manganese electrically from the metal diaphragm, insulating films of consisting of alternate layers of Aluminium trioxide and Silicon dioxide are deposited using electron beam evaporation in presence of oxygen. A cross section of the deposited insulating oxide layers is shown at Fig. 3.16 (*b*).

An overlayer of 2000 Angstrom SiO_2 is deposited for providing protection against atmospheric corrosion. The schematic of the thin film strain gauge system deposited on pressure transducer diaphragm is shown at Fig. 3.17.

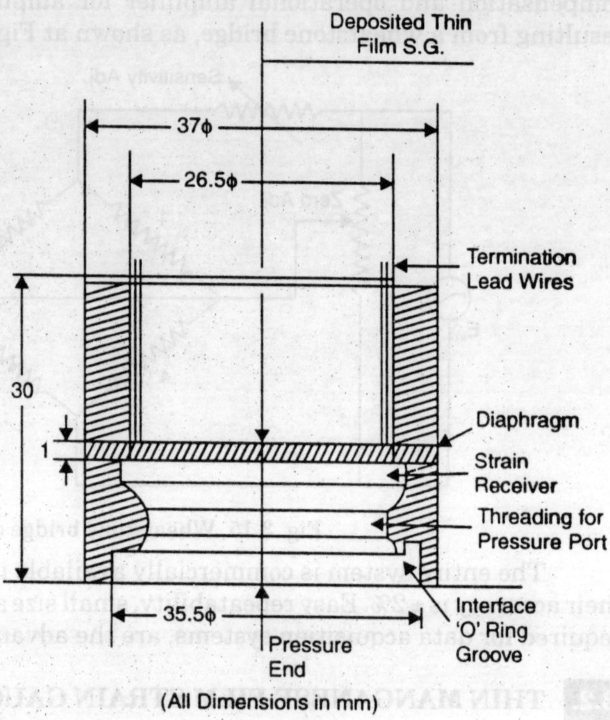

Fig. 3.16. (*a*) A schematic of diaphragm assembly.

The gauge factor which is an index of strain sensitivity was calculated at temperatures – 20°C to 60°C and its variation was found negligible (2.95 ± 0.15).

The output of manganese film pressure transducer was observed to be linear over the entire pressure range of 0—50 bar. It also exhibits good stability over a wide range of time and temperature.

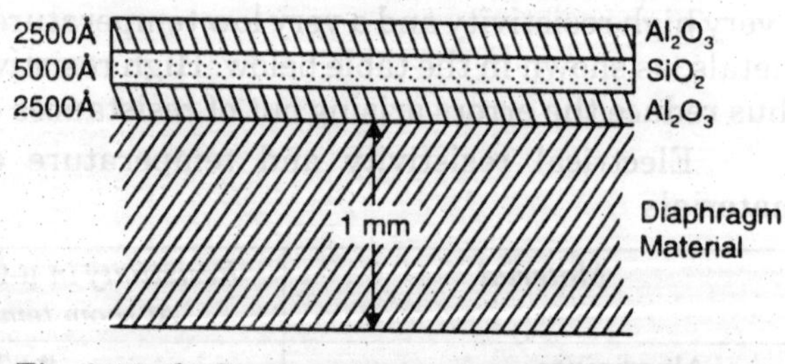

Fig. 3.16. (*b*) Cross sectional view of the insulating oxide layers on a pressure transducer diaphragm.

The main reason why in recent years foil type strain gauges are being replaced by thin film gauges in pressure transducers is that the latter can be directly deposited on the transducer diaphragm, thus eliminating the need to use an adhesive material and the resulting errors that come along with the use of adhesives. Thin film gauges can be deposited on the diaphragm by simple vacuum evaporation techniques or sputtering. It appears promising that thin film pressure transducer can be made to have better characteristics in comparison with conventional transducers in respect of their behaviour, *viz.* linearity, hysteresis, output, stability, repeatability and temperature variation.

While discussing the behaviour of four strain gauges deposited radially and tangentially over the pressure transducer diaphragm, it was observed that the response was linear over a limited pressure range. This linearity applies to very small deflections only. While encountering larger deflections, the behaviour is inherently non linear. As a general rule, the deflection of the

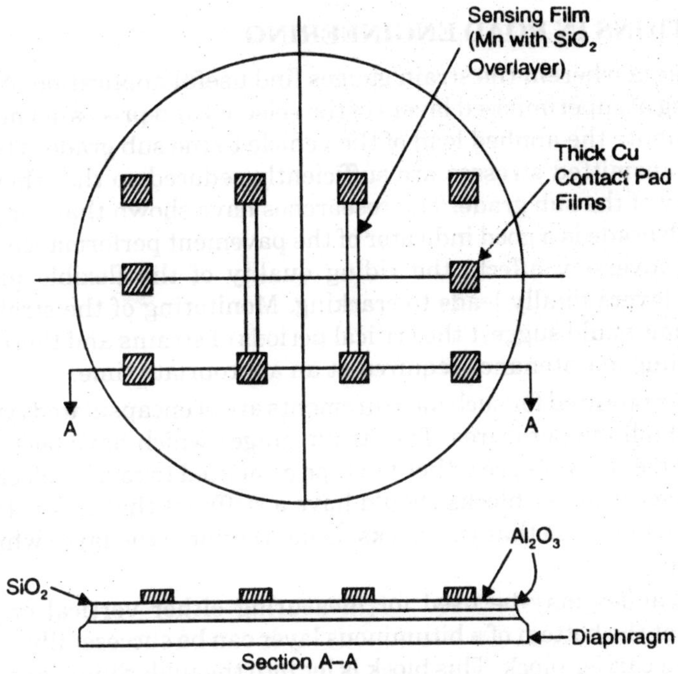

Fig. 3.17. Schematic of thin film strain gauge system on the pressure transducer diaphragm.

diaphragm at the center must not be greater than the diaphragm thickness ; and for linearity of the order of 0.3%, the defection should be limited to $\frac{1}{4}$ of the diaphragm thickness. The general formula for diaphragm deflection, based upon small deflection theory is :

$$d = \frac{3PR^4\,(1-\mu^2)}{16t^3E}$$

where d is the centre deflection in mm.

Frequency Response

In order to faithfully respond to dynamic pressures, it is necessary that the resonant frequency of the diaphragm be considerably greater than the highest applied frequency. Depending strongly on the degree of damping in the diaphragm strain gauge assembly and in the fluid in contact with the diaphragm, the resonant frequency should be three to five times as high as the highest applied frequency. The subject of proper design for accurate dynamic response is too complex.

However, for transducers subject to high frequencies, careful consideration must be given to the frequency response both in terms of amplitude and phase shift.

The undamped resonant frequency of a rigidly clamped diaphragm can be expressed as :

$$f_n = \frac{0.469t}{R^2}\left[\frac{gE}{\omega(1-\mu^2)}\right]^{1/2}$$

where g is the acceleration due to gravity, ω is the specific weight of the diaphragm material and $\omega/g = d$ Kg/cm^3, the density of the material.

3.3. APPLICATIONS IN ROAD ENGINEERING

This is one area wherein the strain gauges find useful application. A road pavement is a structure consisting of superimposed layers of the selected and processed material, the function of which is to distribute the applied load of the vehicle to the sub grade. The ultimate aim is to ensure that the transmitted stresses are sufficiently reduced so that they do not exceed the supporting capacity of the sub grade. The researches have shown that vertical stress or strain at the top of the sub grade is a good indicator of the pavement performance. Strain in the upper bituminous layer adversely affects the riding quality of the flexible pavement. Excessive straining at these layers finally leads to cracking. Monitoring of the strains produced over a sufficiently long time would suggest the critical periods of strains and thereby providing useful information regarding maintenance required at an appropriate time.

Strain gauges required for such measurements are of encapsulated type with the capacity to withstand very high temperatures. The strain gauges which have been used in bitumen or cement bound pavement layers are either foil type cemented to carrier blocks or directly placed on an exposed surface. Carrier blocks should have a stiffness that is less than the material in which they are to be placed so that the blocks do not reinforce the layer which would cause low strains to be recorded.

Foil strain gauges may be used for measuring either vertical or horizontal strains. Horizontal strains at the bottom of a bituminous layer can be successfully measured by a strain gauge cemented to a carrier block. This block is let into the underlying layer so that the surface carrying the gauge is flush with the top of this layer.

Vertical strain gauges have to be placed on carrier blocks and good results are obtained by sandwiching the gauge between two blocks for protection and then placing the block on the surface before paving. This surface can either be the underlying layer or an intermediate layer of bitumen.

Strip transducer of the type shown in Fig. 3.18 is very often used in monitoring of strains during laying of roads. It consists of square steel bars connected by a thin strip of Aluminium that carries a foil strain gauge on either side. Protection of the gauges can be provided by moulding polyethylene around each gauge before installation. The transducer is laid on the surface below the bituminous or the cement bound layer and paving takes place over that layer so that a measure of the horizontal strain at the bottom of the layer may subsequently be obtained. The device may alternatively be cast into a block of the bound material that is then placed face down before paving.

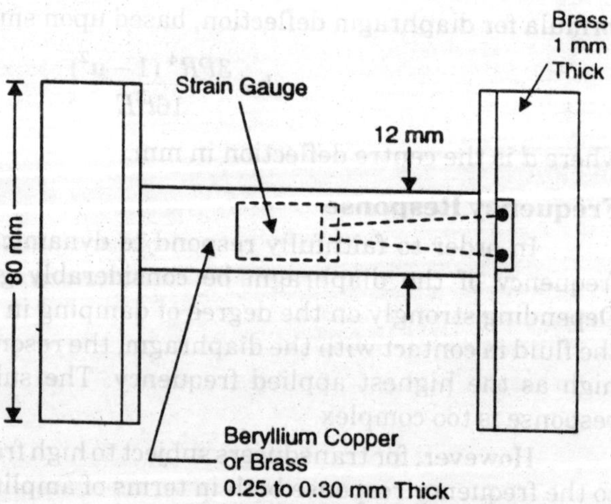

Fig. 3.18. Strip transducer for road engineering.

However, the basic requirement of the strain transducer of any type used in road engineering would be :

(*i*) It must offer restraint to the movement of the material in the direction in which the strain is to be measured, (*ii*) It must measure both tensile and compressive strains and

(*iii*) It must be capable of measuring both small and dynamic strains and large long term strains.

3.4. USE OF STRAIN GAUGE TRANSDUCERS FOR MEASURING AERODYNAMIC FORCES IN WIND TUNNEL BALANCES

Due to their small size and weight, wide frequency response (from DC to over 50 KHz), excellent linearity over a wide range of strain, strain gauge type transducers are preferred in the field of aerodynamics and wind tunnel modelling and balances. Being passive sensors, insensitive to supply voltage and frequency, these can be used both with DC and AC excitation. Their low cost and possibility of being adopted for remote measurements are other advantages. The important parameters for gauge selection for this particular application are :

Gauge Length. It means active or strain sensitive length. The end loops and solder tabs are excluded for this because of their relatively large area and low resistance. Strain measurements are made usually at the maximum peak. Since the maximum peak is restricted to a very small region, the strain gauge tends to integrate or average the strain over the area covered by the grid. Strain distribution is always less than the maximum (Fig. 3.19). Generally, strain gauges less than 3 mm gauge length are used.

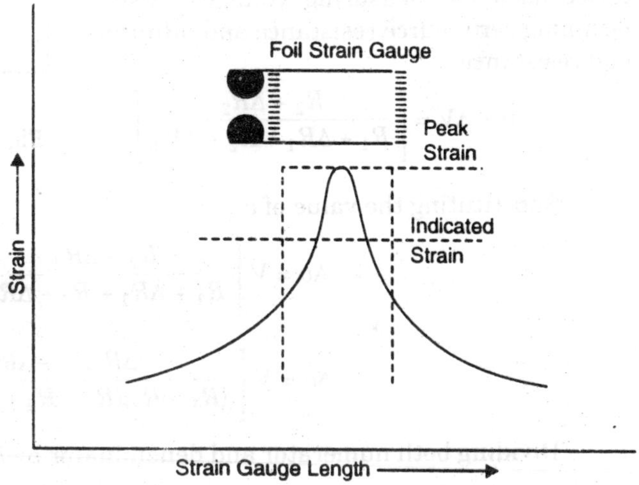

Fig. 3.19. Strain distribution with strain gauge length.

Gauge Pattern (grid geometry). For a gauge length 1.5 mm to 3 mm, grid width of wide patterns are preferred over narrow grid designs because the larger grid area provides better power handling capability. In most of the internal balancing, gauge locations are such that two gauges are installed in close proximity to each other in one or more locations.

In such cases the multiple grid patterns are offered in more than one solder tab configuration. Polymide based gauges are commonly used as these patterns are rugged and flexible and easy to handle. Constantan is also the most popular for transducer gauges because it possesses exceptional linearity over a wide strain range and is readily available in an extensive array of grid configurations. It is also readily solderable and can be subjected to precise temperature compensation when used in conventional bridge circuits.

Gauge Resistance. In wind tunnel applications, many of the mechanical design parameters are often largely fixed by the nature of application. To obtain higher output signal with less power dissipation, high resistance gauges with increased excitation voltage are desirable. Gauges with 120 and 350 ohms are commonly used. However, now 1000 and 5000 ohms gauges in much smaller sizes (1.3×1.5 mm^2) are available and employed.

Self Temperature Compensation. Full bridge configuration is generally used in wind tunnel balances. This provides inherent compensation for thermal output (*i.e.* the apparent strain) if all gauges have identical characteristics and sense identical temperature. Since these conditions can never be precisely satisfied, best accuracy is achieved by selecting self temperature compensation, number of gauges to match the thermal expansion coefficient of the element.

3.5. ELECTRICAL MEASURING CIRCUITS USED WITH STRAIN GAUGES

3.5.1. Potentiometer or Ballast Circuit

This consists of a DC or AC voltage applied across the resistance combination R_1 and R_2 in which either one or both can be strain gauges. The voltage v that appears across R_2, will be given as

$$v = V\frac{R_2}{R_1 + R_2}$$

Assuming a high input impedance device used for measuring voltages. Also, assuming zero source resistance and infinite load resistance.

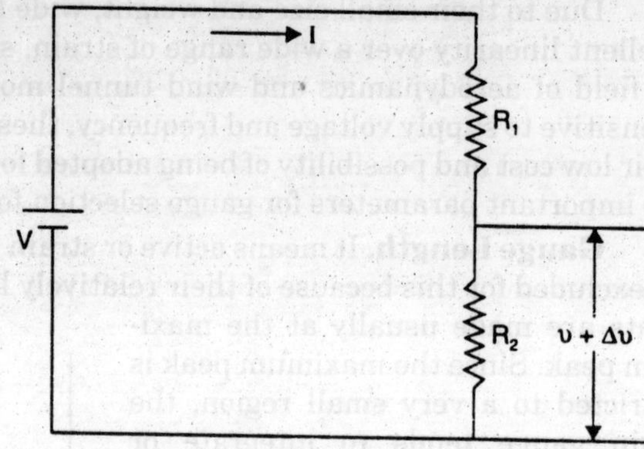

Fig. 3.20. Potentiometric measuring circuit.

$$v + \Delta V = \left[\frac{R_2 + \Delta R_2}{R_1 + \Delta R_1 + R_2 + \Delta R_2}\right]$$

Substituting the value of v :

$$\Delta v = V\left[\frac{R_2 + \Delta R_2}{R_1 + \Delta R_1 + R_2 + \Delta R_2} - \frac{R_2}{R_1 + R_2}\right]$$

$$\Delta v = V\left[\frac{\Delta R_2 R_1 - \Delta R_1 R_2}{(R_1 + R_2)(R_1 + R_2 + \Delta R_1 + \Delta R_2)}\right]$$

Dividing both numerator and denominator by R_1R_2

or
$$\Delta v = V\left[\frac{\frac{\Delta R_2}{R_2} - \frac{\Delta R_1}{R_1}}{\frac{R_1 + R_2}{R_2} + \frac{R_1 + R_2}{R_1} + \frac{\Delta R_1(R_1 + R_2)}{R_1R_2} + \frac{\Delta R_2(R_1 + R_2)}{R_1R_2}}\right]$$

$$= V\frac{R_1}{R_2}\left[\frac{\frac{\Delta R_2}{R_2} - \frac{\Delta R_1}{R_1}}{\left(1 + \frac{R_1}{R_2}\right)^2\left(1 + \frac{\frac{\Delta R_1}{R_2} - \frac{\Delta R_2}{R_2}}{(1 + R_1/R_2)}\right)}\right]$$

Substituting $R_1/R_2 = R$ this becomes :

$$\Delta v = V\left[\frac{R}{(1 + R)^2}\left(\frac{\Delta R_2}{R_2} - \frac{\Delta R_1}{R_1}\right)\right](1 - N)$$

where N is the term accounting for non-linearity, given by

$$N = \left[1 - \frac{1}{1 + \dfrac{1}{(1+R)}\left(\dfrac{R\Delta R_1}{R_1} + \dfrac{\Delta R_2}{R_2}\right)} \right]$$

If R_2 is the only active strain gauge and R_1 is just a fixed resistor, then by neglecting the non-linearity term N, the above expression becomes

$$\Delta v = \frac{R}{(1+R)^2}\frac{\Delta R_2}{R_2} V$$

But since $\Delta R_2/R_2$ = Gauge factor × Strain = Ge

therefore $$\Delta v = \frac{R}{(1+R)^2} Ge \,.\, V$$

The voltage sensitivity $$S_v = \frac{\Delta v}{e} = \frac{R}{(1+R)^2} G \,.\, V$$

When $$R_1 = R_2, S_v = \tfrac{1}{4} G \,.\, V$$

However, the useful range of the potentiometer circuit becomes limited by the non-linearity term N which depends upon the strain level and the resistance ratio R.

From the equation $S_v = \dfrac{R}{(1+R)^2} G \,.\, V$, we rewrite

$$S_v = \frac{V}{(1+R)}\frac{R}{(1+R)} G$$

$$= \frac{V}{(1+R_1/R_2)}\frac{R}{(1+R)} G = IR_2 \,.\, \frac{R}{(1+R)} G$$

where I is the current through R_1 and R_2. Therefore, for a given value of I and gauge resistance, voltage sensitivity increases with R, at the cost of a higher excitation voltage V. This condition also reduces non-linearity. The drawback of the potentiometer arrangement is that it exhibits some initial voltage under zero strain condition which has to be brought to zero value to monitor the small variations of voltage appearing due to strain.

This drawback is overcome in Wheatstone bridge circuits instead of potentiometer.

3.5.2. Wheatstone Bridge Circuits

A typical four arm bridge normally employed for strain measurements is shown at Fig. 3.21.

The potential difference v between points B and D is :

$$v = V_{BA} - V_{DA} = V\left[\frac{R_1}{R_1 + R_2} - \frac{R_4}{R_3 + R_4}\right] = V\frac{[R_1R_3 - R_2R_4]}{(R_1 + R_2)(R_3 + R_4)}$$

This output voltage $v = 0$ when $R_1R_3 = R_2R_4$, when the bridge is balanced. The bridge arms can therefore be adjusted to obtain zero output voltage under no strain condition. There are three main configurations of the bridge circuits :

A-Full Bridge circuit

Where all the four arms contain active strain gauges whose resistance values change by ΔR_1, ΔR_2, ΔR_3 and ΔR_4, under strain conditions.

Then, neglecting second order non-linearity terms and imposing initial balance conditions :

$$v = V\left[\frac{(R_1 + \Delta R_1)(R_3 + \Delta R_3) - (R_2 + \Delta R_2)(R_4 + \Delta R_4)}{(R_1 + \Delta R_1 + R_2 + \Delta R_2)(R_3 + \Delta R_3 + R_4 + \Delta R_4)}\right]$$

$$= V\left[\frac{\Delta R_1}{R_1} - \frac{\Delta R_2}{R_2} + \frac{\Delta R_3}{\Delta R_3} - \frac{\Delta R_4}{R_4}\right]\frac{R_1 R_2}{(R_1 + R_2)^2}$$

$$= \frac{R}{(1+R)^2}\left[\frac{\Delta R_1}{R_1} - \frac{\Delta R_2}{R_2} + \frac{\Delta R_3}{R_3} - \frac{\Delta R_4}{R_4}\right]V$$

where $R = R_1/R_2 = R_4/R_3$.

If G_1, G_2, G_3 and G_4 are the respective gauge factors and e_1, e_2, e_3, e_4 the corresponding strains developed, then

$$v = \frac{R}{(1+R)^2}[G_1e_1 - G_2e_2 + G_3e_3 - G_4e_4]V$$

Normally, in a full bridge configuration, two strain gauges are in tension (R_1 and R_3) and two other in compression (R_2 and R_4) and if $G_1 = G_2 = G_3 = G_4 = G$, the output voltage will be $v = VGe = V(\Delta R_1/R_1)$.

If Poisson's ratio be assumed to be = 0.3, then the output voltage

$$v = \frac{V}{4}[G_1e_1 - G_2(-0.3e_1) + G_1e_1 - G_4(-0.3e_1)]$$

Putting $G_1 = G_2 = G_3 = G_4 = G$ and $e_1 = e$

$$v = 0.65\ VGe = 0.65\ V(\Delta R_1/R_1)$$

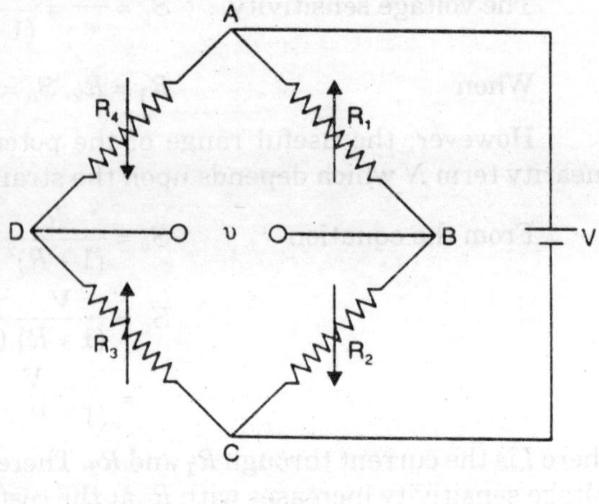

Fig. 3.21. Four arm Wheatstone bridge.

In all the cases, the voltage sensitivity S_v under the conditions of constant voltage excitation, is given by :

$$S_v = \frac{v}{e} = \frac{R}{(1+R)^2}G \cdot V, \text{ which is maximum when}$$

$R = R_1/R_2 = R_3/R_4 = 1$, *i.e.* the equal arm-resistance bridge.

B-Quarter Bridge Configuration

Here, only one arm R_1 contains an active strain gauge while the other three arms contain fixed resistors having a value equal to that of initial unstrained value of R_1

i.e. $R_1 = R_2 = R_3 = R_4$ and $\Delta R_2 = \Delta R_3 = \Delta R_4 = 0$, then

$$v = \frac{R}{(1+R)^2}[G_1e_1 - G_2e_2 + G_3e_3 - G_4e_4]$$

$$= \frac{VGe_1}{4} = \frac{V}{4}\frac{\Delta R_1}{R_1}.$$

C-Half Bridge Configuration

When two arms of the bridge R_1 and R_2 have active strain gauges, one in tension and the other one in compression. The other two arms have fixed resistors of equal value. The output voltage will be given by

$$v = \frac{R}{(1+R)^2}[G_1e_1 - G_2e_2]V$$

If $R_1 = R_2$, $G_1 = G_2 = G$ and $e_1 = e_2 = e$ then

$$\Delta R_1/R_1 = -\Delta R_2/R_2$$

and $$v = VGe/2 = V/2\ (\Delta R_1/R_1)$$

Thus, the voltage sensitivity of the half bridge is twice that of a quarter bridge. If the gauges R_1 and R_2 are placed at mutually perpendicular directions as shown at Fig. 3.22, which is called Poisson's configuration, then :

$$v = \frac{R}{(1+R)^2}[G_1e_1 - G_2(-0.3e_1)]V = 0.325\ \text{VGe}$$

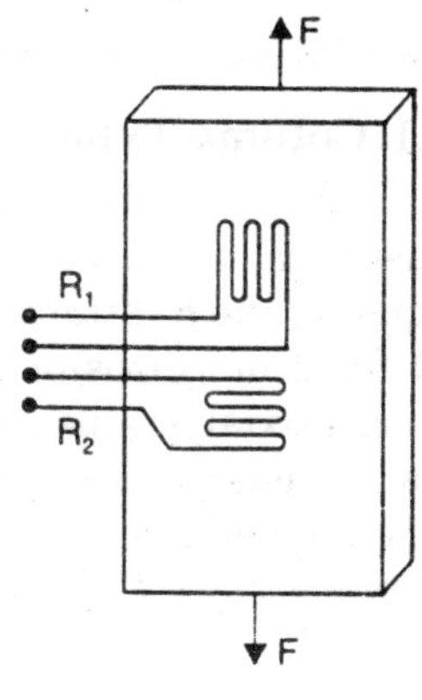

Fig. 3.22. Strain gauges connected in half bridge poisson configuration.

The Wheatstone bridge networks can be used in direct reading mode where null balance is not aimed at and "out of balance voltage" output obtained for an initially balanced bridge is a measure of strain. The voltage measuring device could then be calibrated directly in terms of microstrains. This method is ideally suited for dynamic strain measurements. The only possible sources of error in this arrangement are the possible variations in the excitation voltage and gain instability of the signal amplifier.

The other mode is where the bridge is null balanced every time by variation of resistances in one or more arms.

This latter method is suited for static strain measurements. Since the reading is independent of excitation voltage and signal amplifier gain, the accuracy achievable is very high.

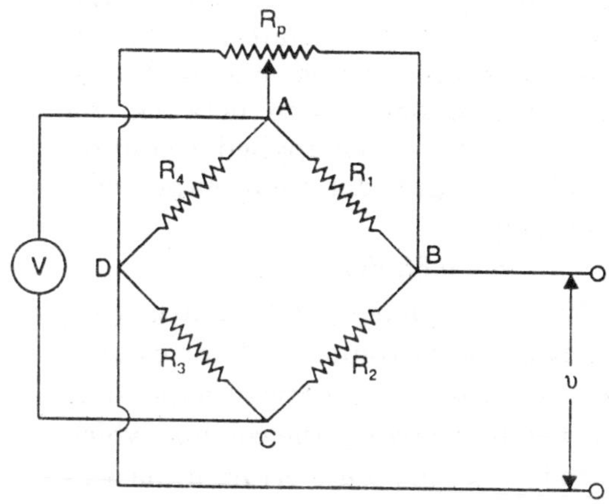

Fig. 3.23. Null method of balancing Wheatstone bridge.

As shown at Fig. 3.23, if R_2 is the strain gauge then the resultant unbalance can be nullified and the bridge brought back to balance by adjusting the potentiometer R_p. The extent to which the potentiometer setting is changed, is calibrated in terms of strain values directly.

3.6. LOAD CELLS (SENSORS FOR MEASURING FORCE)

Load cell is an electromechanical sensor employed to measure static and dynamic forces. It derives its output from the deformation of an elastic member having a high tensile strength and manufactured to very close tolerances, using a suitable steel alloy. The basic design parameters are size, shape, material density, modulus of elasticity, strain sensitivity, deflection and dynamic response. Since strain gauges are used for computing strain, these form a most important device in industrial use employing strain gauges.

The materials chosen for fabrication of load cells should have a linear stress-strain relationship upto a fairly wide elastic strain limit (typically 5000 microstrains), low strain hysteresis and low creep over long periods of cyclic and repeated loadings. The elastic materials suitable are : high carbon steels of Chromium-Molybdenum and hardened stainless steel. Three typical forms of load cells for high, low and very low load ranges will be described.

3.6.1. Column Type

This is the simplest method for measuring unidirectional forces. Strain gauges R_1 and R_3 attached to the column body and fixed on opposite sides are aligned to measure axial strains whereas gauges R_2 and R_4 will measure circumferential strains due to the applied force. The strain values due to an applied force F shown in Fig. 3.24 are :

$$e_1 = e_3 = \sigma/E = F/AE$$

and
$$e_2 = e_4 = -\mu F/AE$$

where σ is the stress developed over cross sectional area A of the column, E is the Young's modulus of elasticity and μ, the Poisson's ratio of the material. Strain measurements are carried out with the standard Wheatstone bridge network. As shown earlier, if V is the applied excitation voltage, G the gauge factor, then, output voltage developed,

$$v = 0.65 \text{ VGe}$$

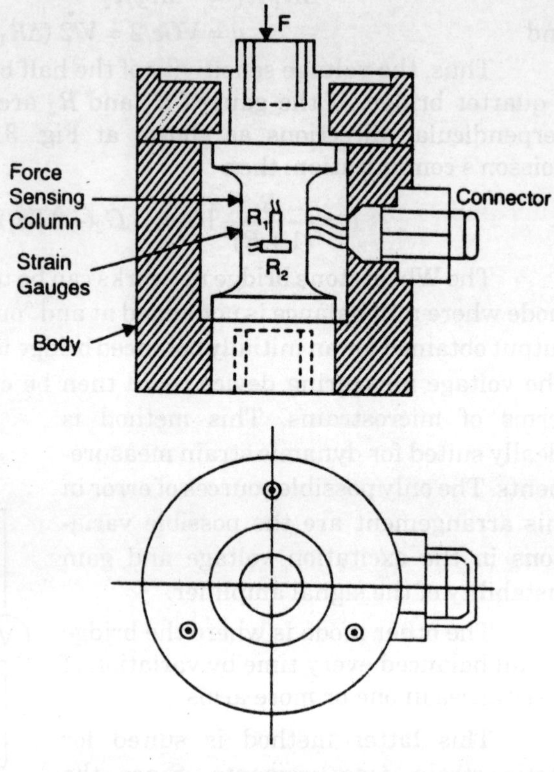

Fig. 3.24. Column type load cell assembly.

While the gauges are mounted symmetrically along the column rod, the column itself should be loaded as centrally as possible to avoid bending forces in the column, which can introduce second order errors if the sensitivities of axial strain gauges are not identical. The load cells are designed in such a way that the stress levels remain well within safe limits (upper limit about 2×10^8 n/m^2 for steel), at the normal load specified. As the nominal load range decreases, the cross sectional area has to be small. One way of doing this is to make the cylindrical column hollow by boring a hole along the length. The lower limit of cross-sectional surface area is that required for the physical bonding of strain gauges. Foil gauges are usually employed. For operation under adverse industrial conditions, the entire cell assembly is hermetically sealed, with the joints welded with electron beam.

3.6.2. Proving Rings

This type of load cell is better suited for medium and low loads and is also known as ring dynamometer. The forces applied across the ring as shown in Fig. 3.25, develop stress along the circumference, in such a way that compressive strains on the inner surfaces and tensile strains along the outer surfaces with the maximum stresses occurring at 90° points of arc in either direction from the point at which the force is applied. The strain gauges R_1, R_2, R_3 and R_4 sense the strain levels. Since both tensile and compressive strains are equal, a relatively higher output can result from the measuring bridge as compared to the axially loaded columns.

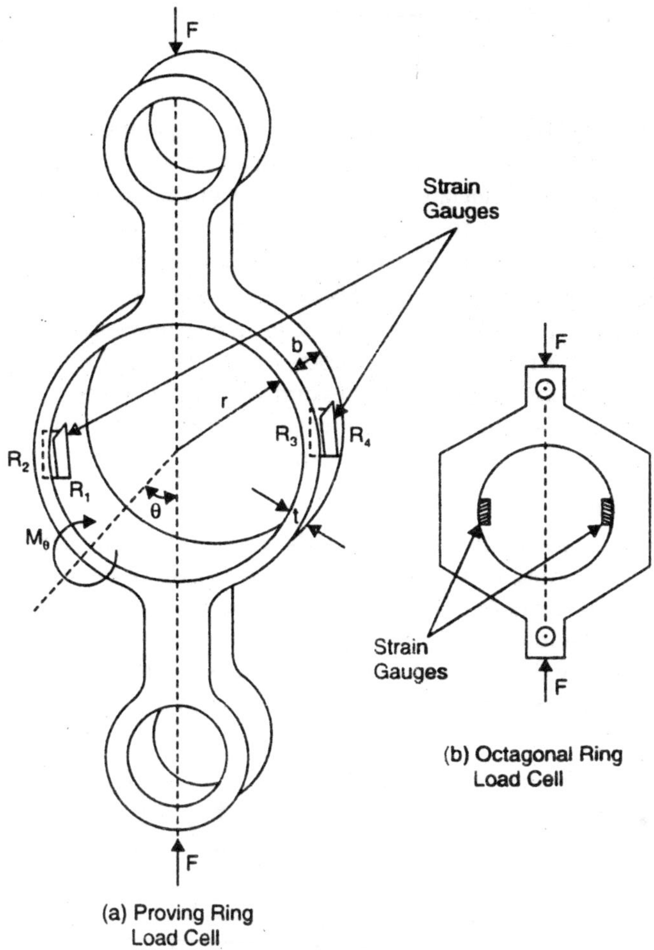

Fig. 3.25. Ring type load cells.

The bending couple for a thin walled ring, M_θ at an angle θ from the direction of the force can be expressed as

$$M_\theta = Fr/2\ (\sin\theta - 2/\pi),\ r \text{ being the mean radius of the ring.}$$

The above relation holds only when no rotation is caused to the top and bottom of the ring. Since strain gauges are bonded at angle θ = 90°, the bending couple

$$M = 0.182\ Fr.$$

The maximum stress σ and the maximum strain e in the thin walled ring are related by

$$\sigma = M_\theta t/2I \quad \text{and} \quad e = \sigma/E$$

Here, t is the wall thickness of the ring and I is the moment of inertia about the axis of bending. $I = bt^3/12$ for rectangular cross section where b is the width of the ring.

Substituting for M and I, the strain magnitude

$$e = \frac{1.092Fr}{Ebt^2},\ \text{where } t << r$$

Therefore, maximum stress at these points

$$\sigma_{max} = Ee = \frac{1.092\, F_{max}\, r}{bt^2}$$

The values thus obtained are theoretical only. The experimental values will, in practice be lower due to the loading couples effect. In order to avoid the tendency of the ring specimen to roll under the influence of any horizontal components of the applied force, the shape is made octagonal so that the strains due to vertical loads only, are predominant. The strains and deflections developed in this case will, of course be lower as compared to the circular ring.

From the expression for maximum stress, it is seen that for a given material, with reduction in nominal load, it becomes necessary either to increase diameter or to decrease thickness of the ring to obtain a particular stress value. Since the increase in diameter and decrease in thickness have their own limits beyond which we can not go, it follows that these limitations to dimensions also limit the application of nominal loads below 50 kg. For still smaller loads, recourse is made to cantilever beams.

3.6.3. Cantilever Beams

This configuration is widely used for loads upto 10 kg. The arrangement is shown at Fig. 3.26. When a force F is applied to the end of the cantilever, a bending moment proportional to the force is developed in the beam. Strain gauges are attached to the top and bottom surfaces of the beam near the fixed end to sense the stresses developed. The tensile strains developed on the top surface are sensed by the strain gauges R_1 and R_3 while the compressive strain developed at the bottom surface area are sensed by the gauges R_2 and R_4.

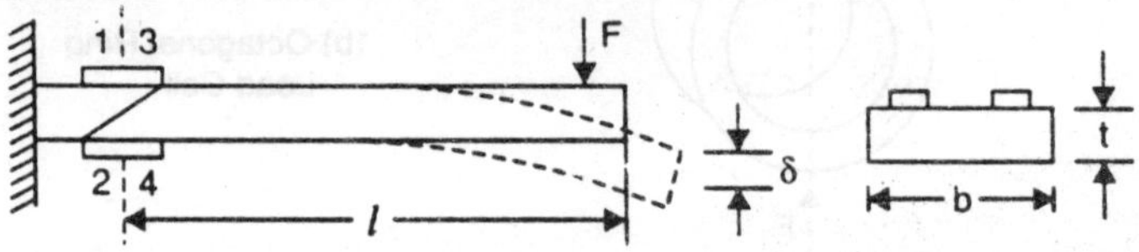

Fig. 3.26. Cantilever beam type load cell.

The maximum deflection due to load will occur at the free end of the beam while the maximum strain will develop at the fixed end. Thus, either deflexion δ or strain e can be measured as the functions of the applied force F, as :

$$e = \frac{6Fl}{Ebt^2}, \text{ at the fixed end}$$

and

$$\delta = \frac{4Fl^3}{Ebt^3}, \text{ at the free end.}$$

Here, again, for getting a particular value of stress for a nominal load, length l is increased or b and t are reduced. For higher loads, either l is decreased or b and t are increased. The former puts a limit on lower nominal load while the latter puts a limit to the higher nominal load. These values are optimised for obtaining best performance. Symmetrical double cantilever arrangements are used for obtaining higher sensitivity.

3.7. STRAIN GAUGE FLOW TRANSDUCER

As an extension to the cantilever type load cell, its use can be made of as a fluid flow transducer. As shown in Fig. 3.27, the free end of the cantilever beam acts as a restrained static vane subjected to a drag force and its net effect can be used as a measurement of fluid velocity.

This is maximum at the tip and decreases towards the fixed end. Hence a coefficient is used to express the force as

$$F = C_d \frac{A\rho V^2}{2}$$

where C_d = Drag coefficient, ρ = Mass density of fluid
V = Velocity of fluid, A = Cross sectional area of pipe

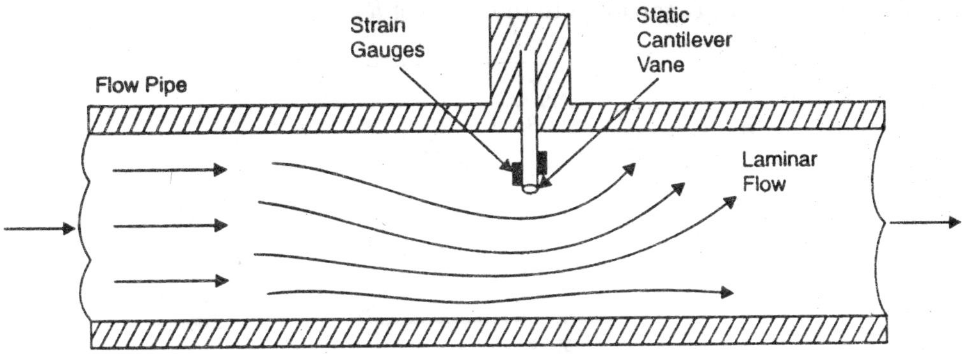

Fig. 3.27. Strain gauge flow tranducer.

In some cases, the orifice plate with many holes is used as a diaphragm and force on it is computed by measuring strain with the help of strain gauges mounted near fixed end of the cantilever, just as in the previous case.

3.8. RESISTANCE TEMPERATURE TRANSDUCERS

This type of temperature sensing again depends on variation of resistance and constitutes one of the most widely used method temperature measurement in the range upto 1000°C. The crystalline structure of metals always contains ions and free electrons existing in equilibrium. As soon as a DC potential is applied, a directional flow of these free electrons results. Their mutual collision and collision with ions in the crystal lattice offers resistance to the current constituted by the moving electrons. As a consequence of rise in temperature on heating, amplitude of oscillation increases thus reducing the mean free path length between collisions. This results in an increase in resistance. However, the crystal structure should not undergo any permanent changes. This condition alongwith the temperature coefficient of resistance and chemical inertness of the metal, decides the range of temperature over which the above phenomenon holds good. The resistance temperature sensor based on the above effect is one of the most accurate device with good reproducibility. In the vicinity of 0°C or 273.15 K, temperature measurement accuracy of 0.0001 K can be attained while within the range upto 1000°C, the accuracy of measurement is between 0.001 K to 0.1 K. Very small temperature differences are also measurable. The only drawback is its large size and relatively sophisticated instrumentation. Platinum, in particular, has excellent stability over a wide temperature range. Platinum resistance thermometer is used to define the international scale of temperature between the boiling point of oxygen – 183.962°C and melting point of Antimony + 630.74°C. Over the range 0—630°C, the resistance variation obeys the polynomial

$$R_T = R_0(1 + aT + bT^2)$$

where, the coefficients $A = 3.983 \times 10^{-3}$ (°C)$^{-1}$ and $B = -0.586 \times 10^{-6}$ (°C)$^{-2}$

T is the temperature in degrees celcius and R_0, the resistance at ice point.

The relatively small temperature coefficient (0.4% per °C) of Platinum requires very careful attention to the usually considered insignificant sources of error in the resistance measuring bridge. This can be eliminated by using three lead wire bridge circuit as shown in Fig. 3.28. Since the lead R_{L1} and R_{L2} are identical and appear in opposite arms of the bridge, changes in lead resistance have no effect on the balance. A special feature of this bridge circuit is that contact resistances of the slider element on R_x and R_y has no effect on balance. By using two mechanically coupled variable resistors such that $R_x = R_y/2$, it can be shown that the null occurs at a position directly proportional to the value of R_T.

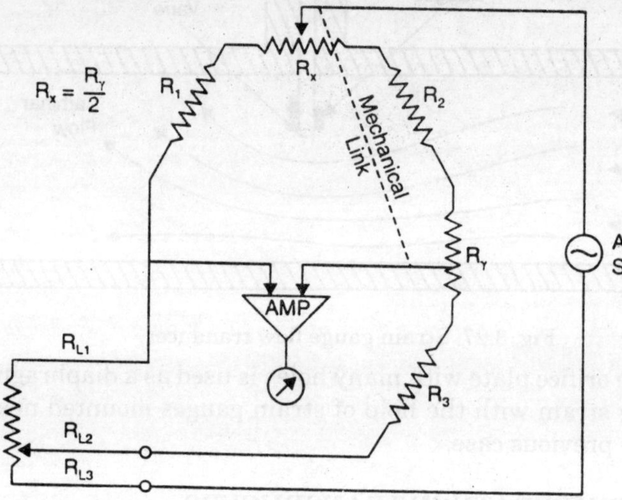

Fig. 3.28. Null bridge for platinum RTD.

If temperature range is not too large and accuracy demanded not too high, a linear indication of temperature can be had by using the bridge in deflection mode, as shown at Fig. 3.29. Here, the arms of the bridge are fixed at values such that the bridge is balanced at lower end of the range, so that 0°C corresponds to zero output. The linearity of this arrangement can be improved by balancing the bridge at the mid-temperature point and also by making $R_1 = R_2$ = about 10 times greater than R_T. Value of R_z in series with the sliding contact of zero adjustment = 100 R_1. Although some sensitivity is lost, it can be compensated by increasing the amplifier gain. Using both these linearity improvements for a Platinum wire sensor, a non linearity of ± 0.5°C over a 0—100°C span can be achieved. An inherent non linearity of the transducer can be compensated by employing a unique linearization technique shown at Fig. 3.30. This is very

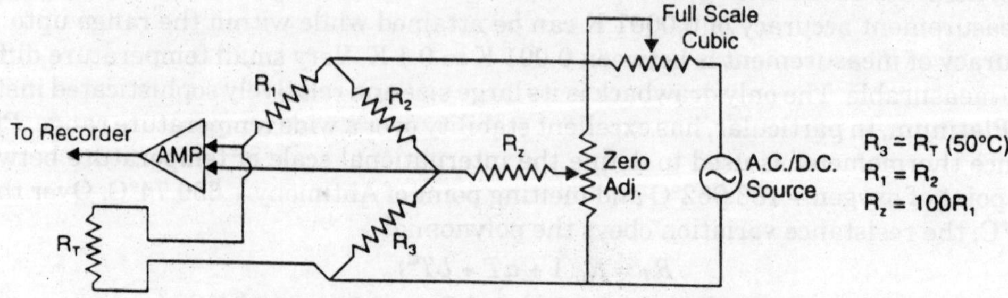

Fig. 3.29. Deflection type bridge for resistance thermometry.

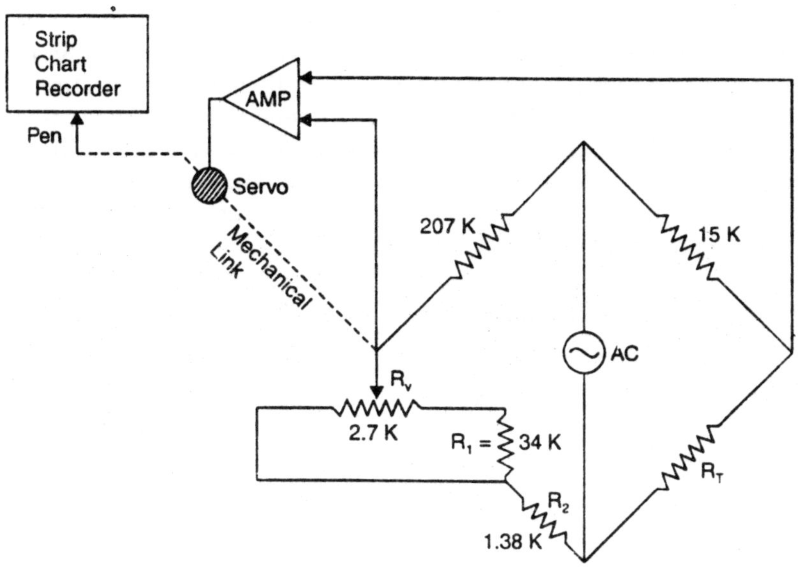

Fig. 3.30. Linearized null bridge using a servo balance (after van der waal and struik).

suitable for high accuracy in temperature measurement over a wide range. Here, the network $R_v - R_1 - R_2$ generates a bridge arm resistance that varies non-linearly with the angular rotation of the multiturn potentiometer in such a way that it compensates for the platinum resistance sensor non linearity. The bridge is of null balance type with the bridge unbalance being applied to the input amplifier of a servo-driven strip chart recorder. A mechanical coupling to the balance potentiometer R_v serves to bring the bridge into balance and thereby to produce an output proportional to the angular rotation. It is claimed that apart from any inaccuracy in the sensor itself, the reading accuracy for a temperature range of 0—500°C is better than ± 0.1°C.

For constructing a sensing element, a small tube of Platinum or silver 4 to 10 inch long is given a coating of a ceramic material, the resistance wire is wound over it and the same is finally covered with another coating of ceramic. This is then heated at a high temperature to ensure annealing of resistance winding and later placed at the tip of the probe enclosed in a protective sheath. Top of the probe tube is often covered with a water proof junction box.

Besides Platinum, Copper and Nickel are other two resistive elements available for making a resistance temperature detector (RTD). It is interesting to compare the properties, merits and demerits of these :

Element	*Temperature range*	*Accuracy*	*Merits*	*Drawbacks*
Platinum	– 180° to 800°C	± 0.5°C	High stability wide range long life	Less linear than copper beyond 660°C, relatively large response time of about 15 sec.
Copper	– 150° to 250°C	± 0.25°C	High stability long life, higher accuracy in the ambient temperature range	Limited temperature range
Nick.l	0° to 325°C	± 0.25°C	Long life, high sensitivity	Lesser linearity

3.8.1. Thermistors

Thermistor is an acronym for "Thermal Resistor". These are essentially semi conductors behaving as resistors with a high negative temperature coefficient. Within a temperature range of – 100 to 300°C, thermistors possess high temperature sensitivity which is utilised for precise temperature measurement, control and compensation. These are normally composed of a sintered mixture of metallic oxides of Mn, Ni, Co, Cu or Fe, are available in various sizes (smallest size that of a bead) and each may have a resistance ranging from 0.5 ohm to 0.75 MΩ. The diameter of a thermistor is usually 0.15 mm to 1.25 mm and a resistivity of 10^{-9} to 10^{9} ohms-cm.

If a small voltage is applied, a thermistor initially draws a very small current and ohm's law is obeyed. When the applied voltage is raised, the current flowing through it is large enough to raise the temperature resulting in decrease in resistance ; with decrease in resistance, more current will flow, further increasing the temperature and causing a further fall in resistance. This process will continue till an equilibrium is reached when the heat dissipated equals electrical power applied. The current drawn increases with increase in voltage until a maximum voltage is reached. Beyond this, the current increases with a decrease in voltage and thermistor displays a negative resistance. This self heating characteristic of thermistor provides an entirely new area of use. In a self heated state, thermistor is sensitive to anything that changes the rate at which heat is conducted away from it. Apart from sensing and measuring temperature, the thermistor can also be used for voltage or power level control when the rate of heat removal from it is fixed because the device becomes sensitive to power input in that condition.

The cross section of a bead type thermistor is shown at Fig. 3.31 (*a*). Its resistance R_T at a temperature T can be expressed by the relation

$$R_T = a \, . \, e^{b/T}$$

where a and b are constants determined by the structure of the material

or $$R_{T1} = R_{T2}\, e^{b(1/T_1 - 1/T_2)}$$

where R_{T1} and R_{T2} are resistance at temperature T_1 and T_2 respectively.

By differentiating w.r.t. T, we get the negative temperature coefficient

$$\alpha = \frac{1}{R_T}\frac{dR}{dt} = \frac{-b}{T^2}$$

a typical value of b is 4000 at a temperature T = 25°C. The normal useful temperature range covered by thermistor is – 100°C to 300°C. However, some thermistors containing sintered Al_2O_3 can be extended even upto 1000°C but at a much coarse accuracy.

3.8.2. Linearization

Variation of resistance with temperature as given by the above relation is obviously non-linear. In the measuring circuit shown at Fig. 3.32, voltage v_0 across thermistor given by $V_0 = R_F(I - I_0)$ is measured directly after amplification through the amplifier A. The conductance *vs* temperature characteristic gets linearised with the help of resistor R_s, in series with the thermistor. In order to effectively eliminate any self heating errors, the series circuit is driven from a voltage source of 50 mv only. Current feedback to the input of the operational amplifier creates a virtual earth at the input terminal, thereby enabling the thermistor current to be measured without affecting the voltage across it. If the amplifier impedance is large, the current through the feedback resistor R_F will be equal to the thermistor current minus I_0. Thus, the output voltage will be linearly related to the thermistor current and by virtue of the series combination having a conductance linearly related to the temperature, the output voltage should also vary linearly with temperature. Such a system when operating over a 0—40°C range, should have a maximum deviation of linearity of around 0.15°C.

However, over a limited temperature range, two alternate approaches for achieving approximate linearisation can be tried. If the thermistor is fed from a constant current source and voltage across the thermistor is used to indicate the temperature, linearisation can be achieved by shunting the thermistor with an appropriately chosen resistor R_p as shown at Fig. 3.31 (*b*). Alternatively, when the current through the thermistor for a fixed applied emf is used to indicate temperature, the series arrangement of Fig. 3.31 (*c*) may be employed. The objective is to make the point of inflection of the parallel on series combination coincide with the mid scale temperature.

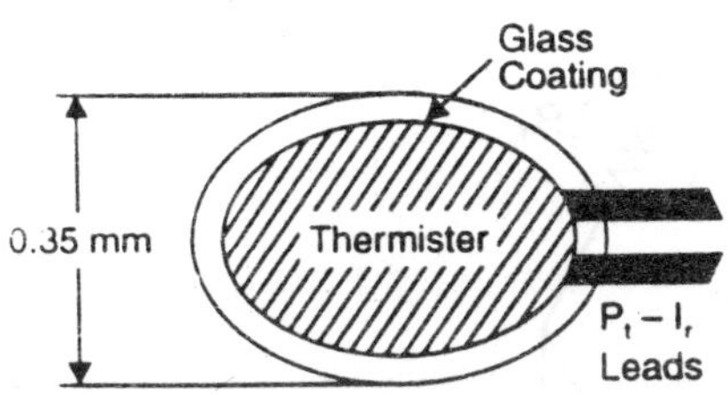

Fig. 3.31. (*a*) A bead thermistor probe.

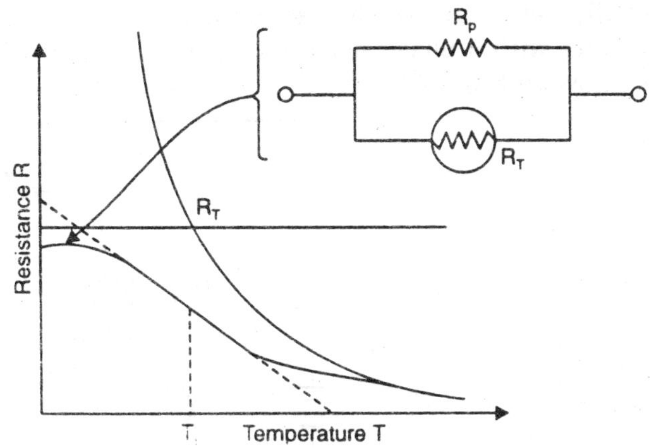

Fig. 3.31. (*b*) Shunt network giving improved resistance-temperature linearity.

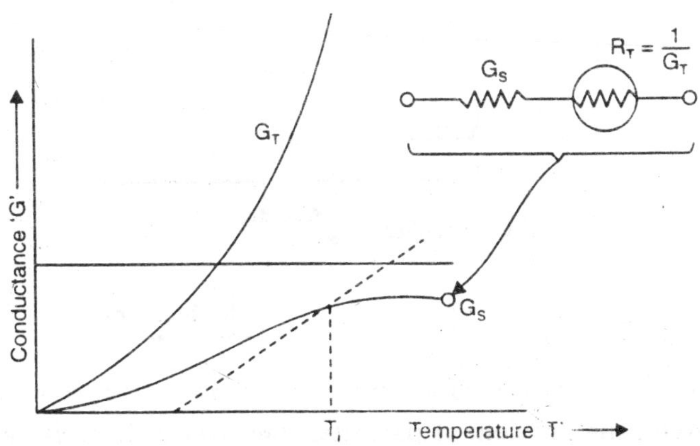

Fig. 3.31. (*c*) Series network giving improved *T Vs G* characteristic.

For better accuracy and high sensitivity, bridge circuits are preferred. By using two thermistors in a differential arrangement, it is possible to achieve very high sensitivity. For this purpose, bead thermistors matched to ± 0.1% of each other at 25°C, are commonly used. If the temperature differences are fairly small, of the order of < 1°C, a DC excited Wheatstone bridge operated in the deflection mode will normally give adequate linearity (better than 1% of full scale output). For a system shown in Fig. 3.33, the stability of DC amplifier is not normally, a significant problem, since even for temperature differences of 0.01°C, the bridge output is still fairly large compared to the equivalent input drift of a good *IC* operational amplifier.

For a still higher sensitivity, it is usually better to use AC excited bridge together with a phase sensitive detector, as shown in Fig. 3.34.

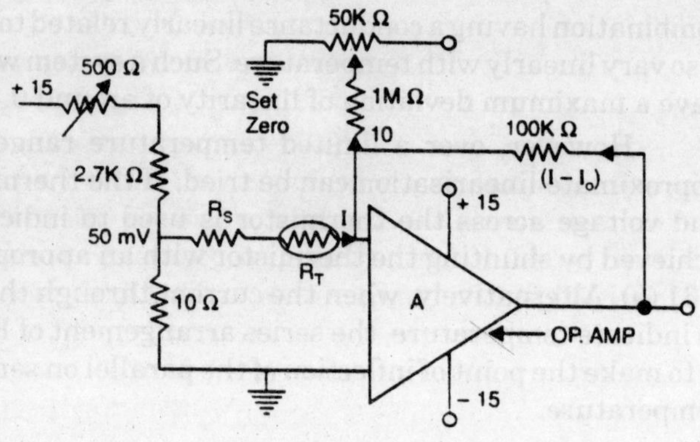

Fig. 3.32. A linearised thermistor circuit for measuring temperature.

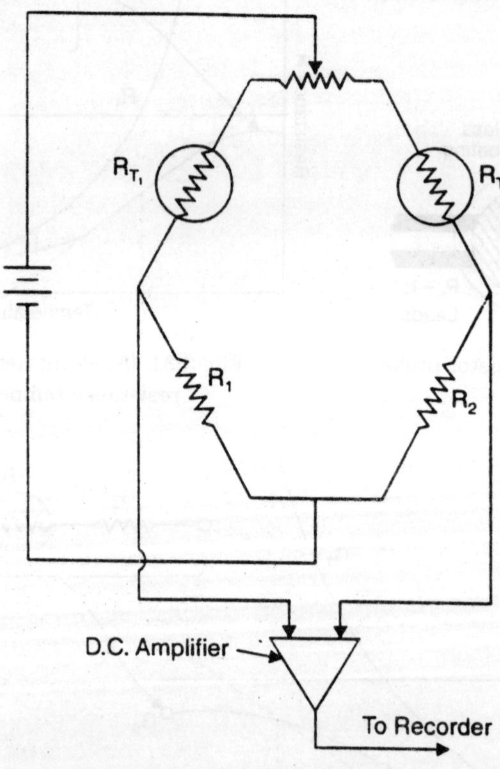

Fig. 3.33. Differential temperaure measuring bridge (Deflection type).

In such a system, the problem of parasitic thermoelectric emfs generated within the bridge is avoided. Since the signal bandwidth required is generally less than 10 Hz, a narrow bandwidth amplifier with its low inherent noise can be used. Also, by using a carrier frequency well away from the low frequency end of the spectrum, the 1/*f* noise generated by both the thermistors and the input stage of the amplifier, is largely avoided. A problem does arise, especially at higher excitation frequencies, from the unbalanced capacitance of the two thermistors. One solution, as shown in Fig. 3.34, is to place a fixed capacitor across one of the thermistors

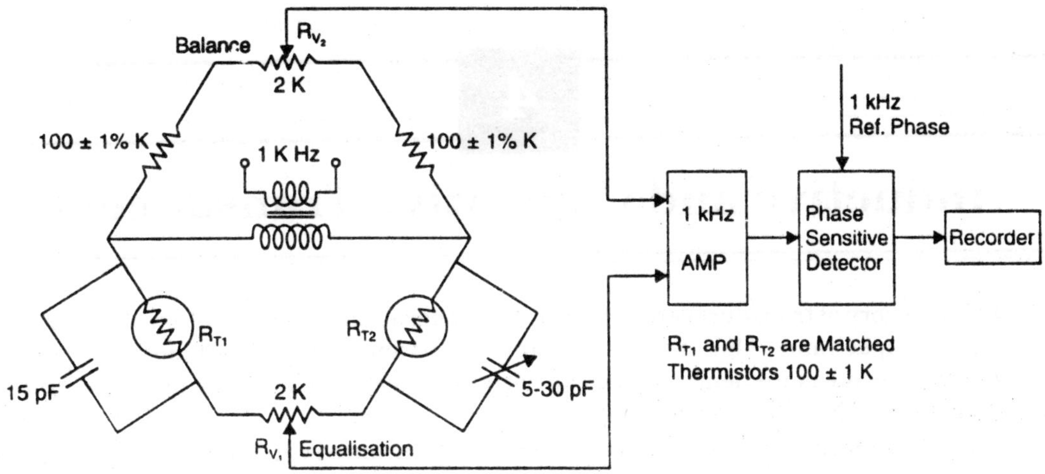

Fig. 3.34. AC excited differential bridge for measuring very small temperature differences.

and a variable one across the other, the latter adjusted to achieve bridge balance when the temperature difference between the thermistors is zero.

BIBLIOGRAPHY

1. DVS Murthy, *Transducers for Instrumentation Systems*, Prentice Hall of India Ltd. 1995.
2. CS Rangan, GR Sarma and VSV Mani, *Instrumentation Devices and Systems*, Tata Mcgraw Hill Pub. Co. Ltd. 1983.
3. Richard SC Cobbold, *Transducers for Biomedical Measurements*, Principles and Applications, John Wiley & Sons 1974.
4. Diamond, JM, *Linearization of Resistance Thermometers and other Transducers*, Rev. Sci. Instrum, 41, 53-60 Jan. 1970.
5. Van der waal, C.W and Struik, LCE, *A Direct Reading Bridge for a Platinum Resistance Thermometer*, J. Sci. Instrum, 2, 143-146 Feb. 1969.
6. Geyling, F.T. and Forst, J.J., *Semiconductor Strain Transducers*, Bell Sys. Tech. J, 39, 705-731 (May 1960).
7. Umesh Sinha, *Electrical and Electronic Measurements and Instrumentation*, Satya Prakashan, New Delhi, Tata India Publications.
8. K. Rajanna, S. Mohan, MM Nayak and N. Gunasekaran, *Thin Film Pressure Transducer with Manganese Film as the strain Gauge*, Sensors and Actuators, A, 24(1990) 35-39.
9. MM Nayak, N. Gunasekaran, K. Rajanna, S. Srinivasulu and S. Mohan, *The Strain Gauge Transducers,* An overview IETE Tech. Rev. 9(2) March-April 1992, 170.
10. PS Krishna Mohan Rao and YS Tiwari, *Strain Transducers in Road Engineering*, IETE Tech. Rev. 9(2) March-April 1992, 167.
11. PN Joshi, *Strain-Gauge Applications in Wind Tunnel Balances*, IETE Tech. Rev. 9(2) March-April 1992, 161.
12. Reznicek I, *Achieving Precision with load cells*, Control and Instrumentation, 5(7) 1973, 40-43.

4

Inductive and Capacitive Transducers

This category of transducers depends for their operation on the variation of self or mutual inductance or capacitance. Out of these, variable inductance type is used mostly for displacement or thickness measurement, though capacitive type can also be made use of for thickness monitoring but with slightly lesser accuracy.

Some other applications wherein variation of capacitance is utilised for indirect measurements are moisture and humidity as also the vibratory motion which converts the capacitive transducer into a microphone.

4.1. INDUCTIVE TRANSDUCERS

4.1.1. Displacement Measurement

The transducers operating on variation of either self or mutual inductance offer considerable sensitivity for displacement or thickness measurement over a limited range. There are certain limitations to the performance of inductors. AC circuits only are to be used with these. Air cored coils when energized, act as sources of alternating magnetic field and develop linkages with the external magnetic fields. Interference with the neighbouring circuits becomes more acute at higher frequencies. As such, proper electrical and magnetic shielding is required. For this reason, iron cored coils are preferred because the magnetic flux then gets contained in iron path and magnetic interference gets minimised.

The simplest type of inductive transducer shown at Fig. 4.1 is essentially a variable reluctance type. It consists of a coil wound on a ferromagnetic core C, with a movable target of iron to which displacement is to be applied. There is no physical contact between the two and

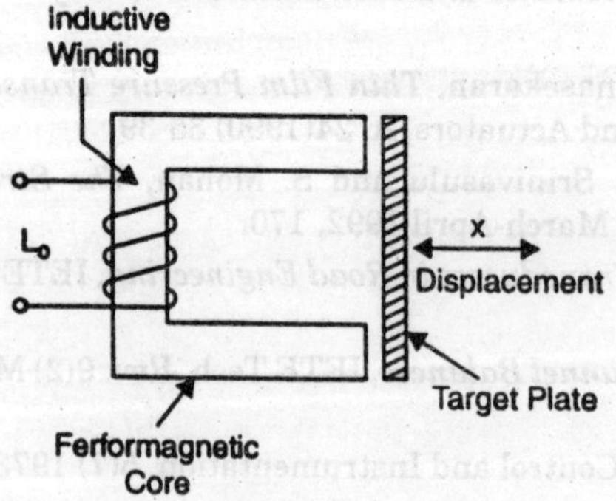

Fig. 4.1. (*a*) Simplest type of inductive transducer

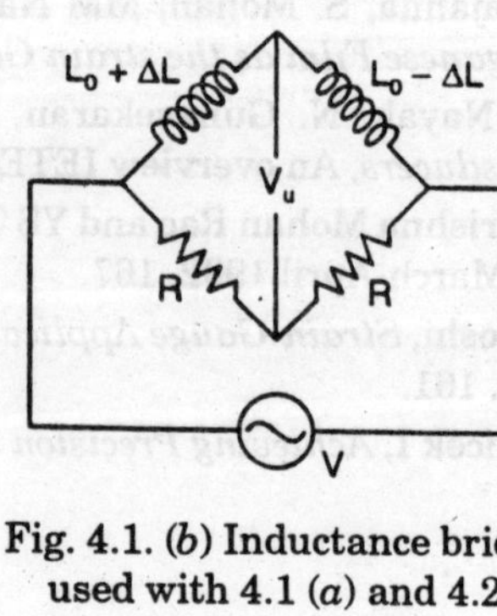

Fig. 4.1. (*b*) Inductance bridge used with 4.1 (*a*) and 4.2.

these are separated by an air gap. The coil inductance depends on the reluctance of the magnetic circuit. It is given by :

$$L = N^2/(R_i + R_a)$$

where N is the number of turns on the coil and R_i, R_a denote the reluctance of iron part and air gap respectively. R_a is given by :

$$R_a = l_a/\mu_o A_a$$

where l_a and A_a indicate length and cross section area of air gap and μ_o is the permeability of the magnetic circuit

Since $\qquad R_a >> R_i, L \approx N^2/R_a \approx N^2\mu_o A_a/l_a \qquad$ ...(4.1)

Therefore $\qquad L \propto l/l_a$

or coil inductance is inversely proportional to l_a, the length of the air gap or in other words, the linear displacement imparted to the iron target T.

This arrangement suffers from the following drawbacks :

(*i*) The initial displacement has to overcome the electromagnetic force of attraction.

(*ii*) The satisfactory linearity between fractional changes in air gap or displacement and the corresponding fractional changes in coil inductance, at any chosen value of air gap can be found only for very small value of $\Delta l_a/l_a$, say within ± 0.01.

The changes in coil inductance L can be measured by an AC bridge network.

Some of the basic drawbacks as indicated above, can be overcome by using two identical cores, with identical coils, so as to constitute a push-pull transducer. The armature or movable core subjected to displacement in the same direction as that of the electromagnetic force, so that the net force on the armature is zero when it is symmetrically between two cores. The coils have some self-inductance when the core is in neutral position. The arrangement is shown in Fig. 4.2.

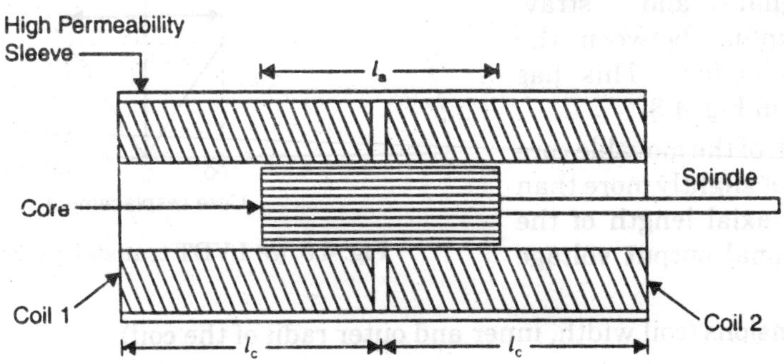

Fig. 4.2. An inductive transducer with push pull arrangement using two identical coils.

The inductance of coil depends on the length of that part of the core, which lies inside the coil. Any movement of the core from initial null position disturbs the equality in values of inductance and results in an unbalance voltage V_u. [Fig. 4.2 (*b*)]. The advantages of this arrangement are :

- Immunity from the effects of external magnetic fields.
- Considerable reduction in the force required to move the core from null position.
- Force required to overcome electromagnetic pull on the armature is the same for movements to either side.
- There is very little magnetic flux that could interfere with the neighbouring circuits.

4.1.2. Linear Variable Differential Transformer (LVDT)

A further refinement is achieved with the addition of another coil, primary, which functions on the basis of mutual inductance M instead of self-inductance L. Two secondaries are connected in series opposition so that the emfs generated or induced in them oppose each other. Two secondaries are axially placed and wound on a cylindrical former. A rod shaped magnetic core is placed centrally inside the coil assembly, thus providing a path for the magnetic flux linking the coils. The displacement to be measured is transferred to the core through suitable linkages. The variable coupling between primary and secondaries connected in series but in phase opposition, produces a resultant output voltage so that it is zero when the core is symmetrically placed.

The entire system is known as a "Linear Variable Differential Transformer (LVDT)". The extremely fine resolution, high accuracy and good stability make the device particularly suited as a short stroke, position measuring device. At symmetrical position, two equal voltages are induced in two coils, in phase opposition to each other. The magnitude of the resultant voltage is zero at this "Null Position". In practice, a small residual voltage is always present even at this null position due to the presence of harmonics in the excitation signal and stray capacitance coupling between the primary and secondary. This has been illustrated in Fig. 4.3.

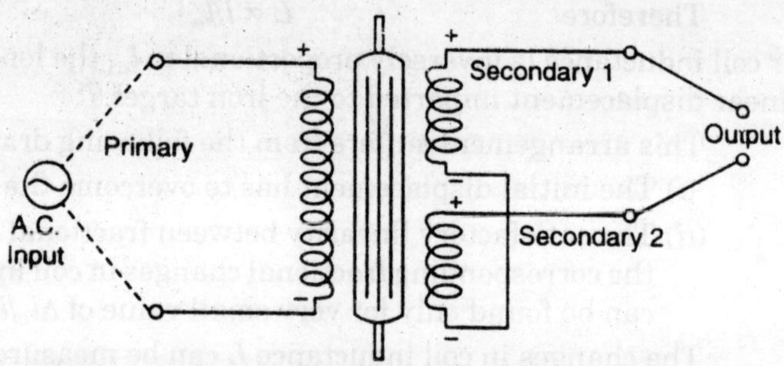

Fig. 4.3. (*a*) LVDT (Secondaries in phase opposition.

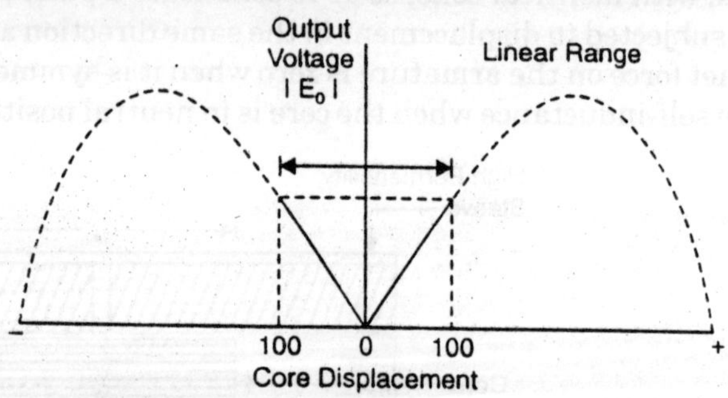

Fig. 4.3. (*b*) LVDT transfer characteristics.

The length of the movable core is, in practice kept slightly more than three times the axial length of the primary. The signal output voltage depends upon :

- Dimensions (coil width, inner and outer radii of the coil)
- Primary current
- Core displacement
- Frequency of excitation voltage

The net induced emf E_0 is given by Atkinson formula, which approximates, the relationship between the coil data and the LVDT output.

$$E_o = \frac{16\pi^3 f I_p n_p n_s \cdot 2bx}{10^9 \ln(r_o/r_i) \cdot 3w}\left(1 - \frac{x^2}{2b^2}\right) \qquad \text{...(4.2)}$$

where f = Excitation frequency

I_p = Excitation current

n_p, n_s = Number of turns on primary and secondary respectively

b, w = Width of primary coil and each secondary coil respectively

r_o, r_i = Inner and outer radii of the coil

x = Core displacement

The non-linearity term $x^2/2b^2$ depends on the length of the primary winding.

For best linearity, core displacement $x << b$. Design is optimised for lowest null voltage, best linearity and appropriate size. Optimum ratio r_o/r_i is usually kept between 2 and 8.

As mentioned above, when the core is symmetrically placed with respect to secondary coils, resultant output voltage induced should be zero but in practice, a small residual voltage is always present due to the presence of stray capacitance between primary and secondary and harmonics in excitation voltage signal. This small residual voltage can be reduced either, by using a high resistance potentiometer and adjustable capacitor or by using an excitation source with a centre tapping and one end of secondary terminals tied together, and grounded. The former arrangement when used together with the latter, balances the secondary impedance with respect to its centre tap. This is shown in Fig. 4.4.

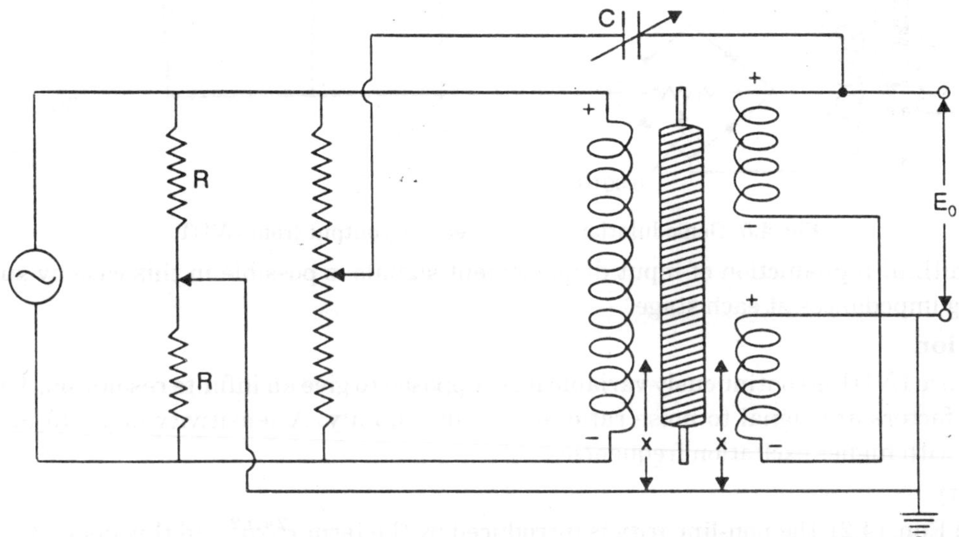

Fig. 4.4. Arrangement for reducing residual null voltage in LVDT.

For greater sensitivity, the number of turns on secondary should be as large as possible and since it is to be connected to a high impedance load and amplifiers, it is to be wound with finer wire. Typical sensitivity is usually about 1-2 mv per volt of excitation voltage per 0.01 mm of displacement, at an excitation frequency of 2-5 KHz. The excitation voltage is limited by the maximum primary current that can be allowed without causing thermal drift in the transducer performance.

At very low frequencies (60 Hz – 1 KHz) and at very high frequencies (10-20 KHz), the phase difference between the primary input and secondary output voltages is high (± 90°). For every LVDT, there is a particular frequency somewhere in the middle range, where the phase difference would be zero. It is always desirable to choose a frequency close to that for better sensitivity and proper demodulation.

Dynamic Response

This is also limited mainly by excitation frequency, which has to be at least ten times higher than the maximum frequency of measurement to distinguish the signal from carrier. At higher dynamic frequencies (typically 2 KHz), the effect of mechanical loading due to core weight, is to be taken into account.

The output signal is an amplitude-modulated sine curve. Demodulation is effected by phase sensitive detectors. If the induced emf of each secondary is separately available, a simple demodulation circuit using two full wave rectifiers and a cascaded RC filter is used. (Fig. 4.5).

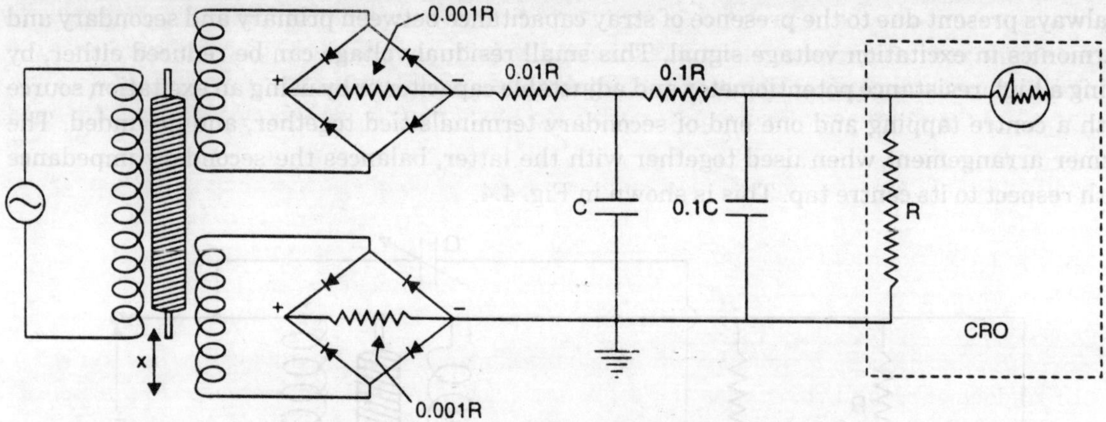

Fig. 4.5. Demodulation circuit used with output from LVDT.

Faithful reproduction of input displacement signals is possible in this case by suitably matching impedances at each stage.

Resolution

Since LVDT is continuously variable it is supposed to give an infinite resolution. The only limiting factors are signal to noise ratio and input stability. A sensitivity of 10-40 mv/μ_m is possible with higher excitation frequencies.

Linearity

In Eqn. (4.2), the non-linearity is introduced by the term $x^2/2b^2$ and this depends largely on the length or width of primary coiling. Standard LVDTs, designed for displacement range upto 5 mm have a good linearity upto 0.1% FSO. Long range LVDTs (range upto 300 mm or more), which are compensated for higher range, can give linearity upto 1-2% FRO.

Full-scale displacements of 0.01-25 cm, with a linearity of ± 0.25% are available as a standard marketed item.

With careful design of circuitry, 0.1% of full scale can be measured, enabling displacements of less than 0.1 μ_m to be detected. The core weight is less than 0.28 for the 0.01 cm unit, and increases to around 10 g for the 25 cm unit. Since the core has a small radial clearance, no frictional forces need be involved in its displacement, provided it can be held sufficiently rigidly by the source of motion. The only axial forces that do arise are caused by the core weight and a small emf that tends to restore to its null position. Also present is a small radially directed electromagnetic force that may necessitate providing an insulating core guide to prevent the core touching the bore.

4.1.3. Other Variable Inductance Transducers

A variable inductance transducer capable of better overall stability and linearity, making use of a balanced coil arrangement is somewhat similar to the one shown in Fig. 4.2. In a modified arrangement, each coil controls the frequency of a separate oscillator, as illustrated in Fig. 4.6. For zero displacement, the oscillator frequencies f_1 and f_2, differ slightly, so that the frequency mixer delivers an output $f_3 = f_1 - f_2$ which is much less than either f_1 or f_2. When the core is moved towards L_1, the inductance of L_1 increases and that of L_2 decreases, causing the oscillator frequencies to change to $[f_1 - \Delta f + k(\Delta f)^2]$ and $[f_2 - \Delta f + k(\Delta f)^2]$. Since the mixer and filter take the difference in frequencies, the second order non-linearities get cancelled, yielding a frequency $(f_3 - 2\Delta f)$. By means of a frequency to voltage converter, the change in output frequency $2\Delta f$ can be converted into an analog output voltage signal. Alternatively, by using a digital counter, the frequency change can be presented in digital form.

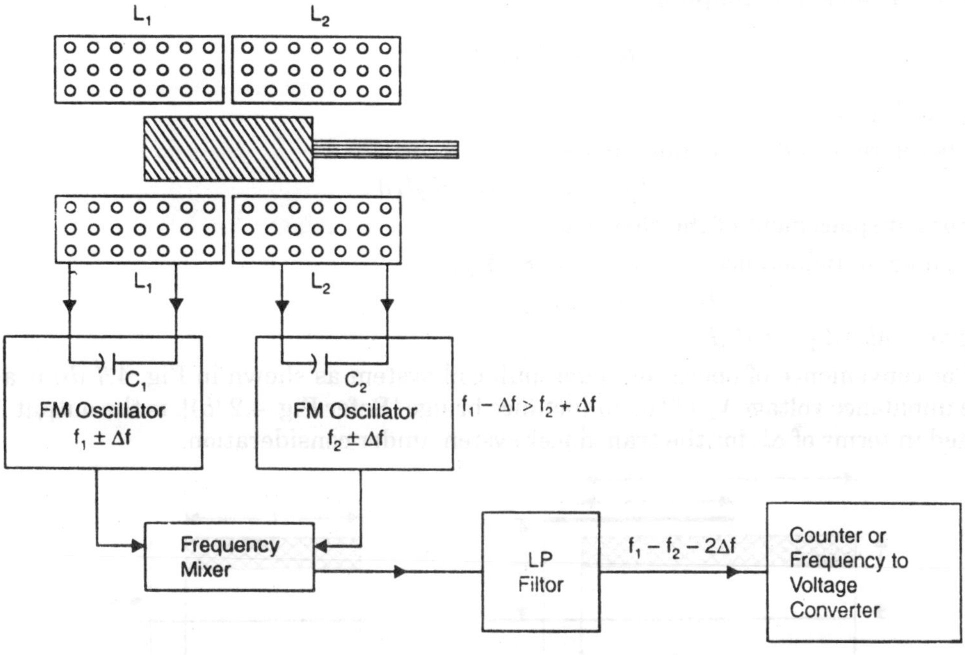

Fig. 4.6. A two coil variable inductance transducer system.

4.1.4. Eddy Current Type Inductive Transducer

The inductive transducers described above are based on the movement of either a core or a coil, which though of very small mass, could under certain circumstances disturb the measured quantity. One inductance scheme that enables a contactless measurement to be made, is based on the change in eddy currents induced in a conducting surface by a coil. Thus, this type of transducers can be described basically by a coil that produces a magnetic field at right angles to the surface of an electrically conducting plate.

When the coil carries alternating current, the eddy currents produced in the conducting surface act like a shorted secondary and hence the effective self inductance of the primary coil is reduced. Conducting sheath is normally of a non magnetic material but it can also be of magnetic one. The transducer can be used as a proximity meter to sense the approach of a metallic object near the coil and is calibrated to indicate closeness for each situation. The effective change in self-inductance of the coil depends upon the separation, thickness and electrical conductivity of the material of the surface, on coil dimensions and frequency of AC excitation. The geometric orientation of the surface of the approaching object should be in conformity with the design and calibration. For non magnetic surfaces, the excitation frequency may be given in the RF range. For regularly shaped objects, like a thin walled circular ring or sleeve or hollow tube of copper or any other conducting material, it is possible to work out approximate variation is self inductance of the coil.

The ring or sleeve acts as a single turn secondary. The effective inductance of a coil coupled with a secondary is given by

$$L_1' = L_1(1 - K^2)$$

where L_1 = Self inductance of the primary, and

K = Coefficient of coupling between primary and secondary.

If l_2 denotes the sleeve length and r_2 its radius over the primary coil of length l_1 and radius r_1, then the coefficient of coupling

$$K = \frac{r_1}{r_2}(l_2/l_1)^{1/2}$$

as shown at Fig. 4.7 (a).

Therefore the effective inductance

$$L_1' = [1 - (r_1/r_2)^2 (l_2/l_1)]$$

For a displacement of the sleeve Δl_2,

Change in inductance $\Delta L_1' = -L_1K^2 \Delta l_2/l_2$

or $L_1K^2 \Delta l_2 = \Delta L_1' l_2$

which gives $\Delta l_2/\Delta L_1' = l_2/L_1K^2$

For convenience of operation, push-pull coil system as shown in Fig. 4.7 (b) is adopted and the unbalance voltage V_u of the inductance bridge [Refer Fig. 4.2 (b)], is the output. This is calibrated in terms of Δl_2 for the transducer system under consideration.

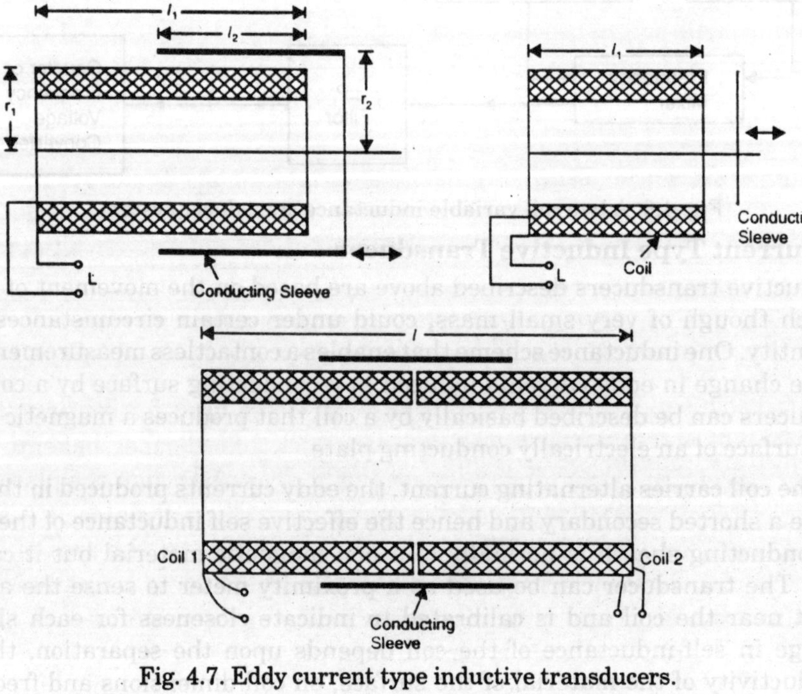

Fig. 4.7. Eddy current type inductive transducers.

The system has a low sensitivity but can cover a large range. Because of the extensive spread of its magnetic field in and around the coil, this type of transducer finds wide use.

4.2. CAPACITIVE TRANSDUCERS

The dependance of capacitance on plate area, their separation, and permittivity of the medium between the plates of a capacitor have all been taken as a means of converting a displacement, motion or force to an electrical output. The main difficulty of working with capacitive transducers is the relatively small change in capacitance with these physical parameters. Special circuit techniques are therefore necessary to register the small capacitance variations accurately, avoiding spurious stray capacitance effects. Nevertheless, these offer excellent accuracy and sensitivities of the order of 10^{-12} cm. The capacitance microphones for example make use of variation of capacitance with motion, between a fixed plate and diaphragm acting under sound pressure. These are in fact, considered to have best performance characteristics of all other types. Capacitive transducers are ideally suited as non contact type dynamic ones, in studies of vibration in light structures such as thin walls and diaphragms. Their non-contact property is desirable when a mechanical displacement is required to be measured without causing any additional mechanical loading of the vibratory object. Capacitive transducers also exhibit high stability, good linearity and working temperature range and above all, compactness.

The most common form is a parallel plate capacitor with a variable gap, '*d*' between the plates or variable area of one of the plates. The third possibility is that of change in dielectric constant of the medium separating the two plates. Fig. 4.8 illustrates a basic type, with a

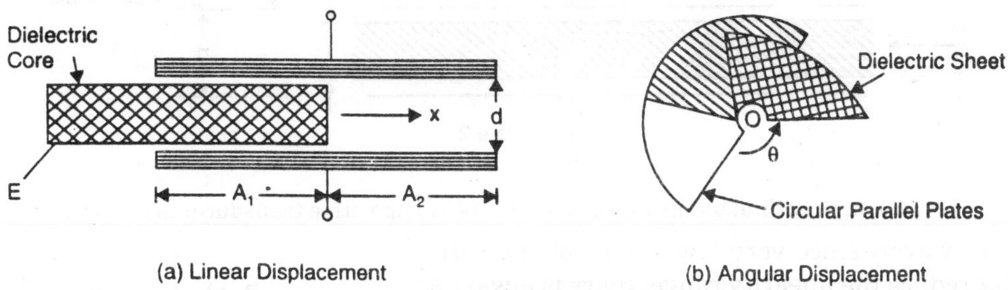

Fig. 4.8. Capacitive transducer using movable dielectric core.

movable dielectric core between the two parallel plates for measuring linear displacement and a movable dielectric sheet for measuring angular displacement. However, mass of the moving system and the electrostatic force on the dielectric limit the dynamic response of such transducers at low frequencies.

4.2.1. Thickness Monitoring

In Fig. 4.9 is shown an arrangement which can be used for continuous non-contact monitoring of thickness of their insulating sheets *e.g.* glass, mica, paper, bakelite etc. which are under constant production. The thin sheets are made to pass through the plates of a parallel plate capacitor and then to move out through a parallel pair of guard plates. It '*t*' is the thickness of the sheet passing through (this acts as a dielectric medium), *A*, the area of each capacitor plate, *d* the total distance that separates the two plates (including a small air gap on either side), ε_r, the dielectric constant or relative permittivity of the material and ε_o, the absolute permittivity of air or free space, then ε_r remaining constant, capacitance

$$C = \varepsilon_o \varepsilon_r A/d = \varepsilon_o \varepsilon_r A/(d - t) + t$$

where $(d - t)$ is the air gap while 't' is the thickness of the dielectric material.

This gives : $C = \varepsilon_o A/[(d - t) + t/\varepsilon_r]$ (dividing by ε_r)

$$C = \varepsilon_o A/[d - t(1 - l/\varepsilon_r)]$$

An increase value of 't' by a small amount Δt causes a corresponding increase in capacitance C by ΔC.

So that
$$C + \Delta C = \frac{\varepsilon_o A}{[d - (t + \Delta t)\{1 - (1/\varepsilon_r)\}]}$$

Dividing both sides by C and solving :

$$\frac{\Delta C}{C} = \frac{\Delta t N}{t[1 - N(\Delta t/t)]}$$

where N is the sensitivity factor given by

$$N = \frac{\varepsilon_r - 1}{1 + \varepsilon_r (d - t)/t}$$

For $N(\Delta t/t) << 1$, $\Delta C/C \propto \Delta t/t$ or change in capacitance C expressed as a fraction of C is proportional to the corresponding change in thickness 't' expressed as a fraction of 't'.

Sensitivity factor N is maximum for very very small values of the ratio $\varepsilon_r(d - t)/t$ or in other words air gap $(d - t)$ to thickness ratio 't' very small, say of the order < 0.01.

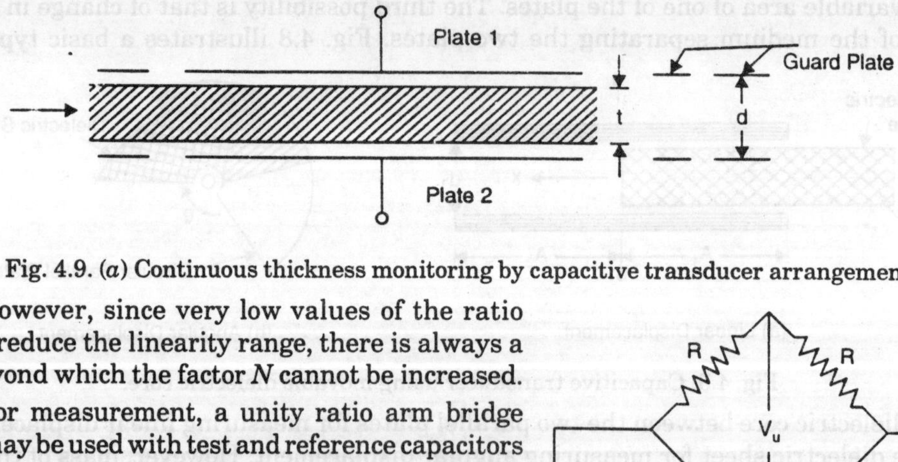

Fig. 4.9. (*a*) Continuous thickness monitoring by capacitive transducer arrangement.

However, since very low values of the ratio $(d - t)/t$ reduce the linearity range, there is always a limit beyond which the factor N cannot be increased.

For measurement, a unity ratio arm bridge circuit may be used with test and reference capacitors in the opposite arms as shown in Fig. 4.9 (*b*). The reference capacitor has the dielectric sheet of the same material but of a constant thickness, inserted between its plates. The unbalance voltage V_u may be calibrated in terms of the values of $\Delta t/t$ so as to make it direct reading. Very small values of $\Delta t/t$ can thus be detected.

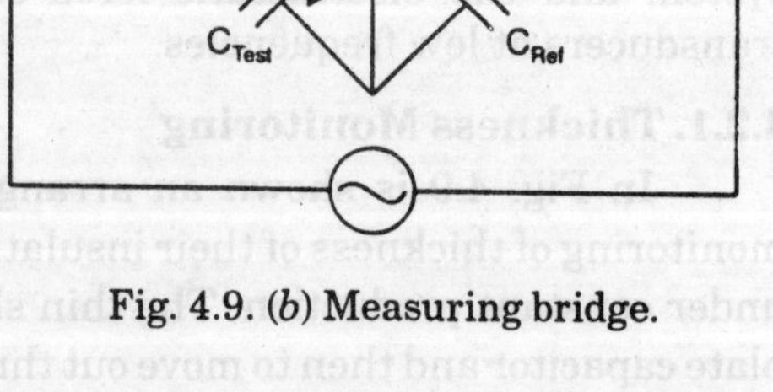

Fig. 4.9. (*b*) Measuring bridge.

4.2.2. Capacitive Displacement Transducers

In a typical arrangement, one of the parallel plates of the capacitor is made movable and subjected to mechanical displacement. If the transducer has a solid insulating material of dielectric constant C, the capacitance $C = \varepsilon_o/(x + t/\varepsilon)$ where x is the displacement and ε_o, the permitivity of air or free space.

If the air gap is increased or decreased by Δx, then C will decrease or increase by :

$$C \mp \Delta C = \frac{\varepsilon_0 A}{x \pm \Delta x + t/\varepsilon}$$

which gives fractional change in capacitance

where $N = \dfrac{1 + t/x}{1 + t/x\varepsilon}$

gives the sensitivity factor.

For small values of $N\Delta x/(x + t)$, percent changes of capacitance are proportional to percent changes in separation between the plates. If solid dielectric material is removed, $N = 1$ obviously, the presence of solid dielectric improves sensitivity by making $N > 1$, but linearity gets reduced. Due to this reason, air dielectric capacitor is commonly used. In that case, the non linearity term

$$\frac{N}{1 \pm N\Delta x/(x + t)}$$

can be neglected and for small changes Δx in the position of the movable plate with respect to an initial value, the capacitance variation ΔC can be written as : $\Delta C/C = -\Delta x/(x + t)$, which is a linear relationship. The minus sign implies that there is an increase in capacitance as the movable plate is moved towards the fixed plate and a decrease as it is moved away.

A linear variable capacitance transducer in a practical application comprises a cylinder moving concentrically inside two other identical cylinders with a constant air gap. With the movement of the central cylinder, the capacitance area of one section increases while the other correspondingly decreases..The resultant change in capacitance is directly proportional to the displacement of the inner cylinder. The main advantages of this device are good stability, and a linearity better than ± 2% over a 25 cm range. This has been shown in Fig. 4.9 (*b*). Such transducers are employed for level measurements in liquid columns, as in the fuel gauge system of an aircraft or level indicator in a chemical plant. The resultant change in capacitance can be converted into a useful electrical signal for the convenience of measurement.

4.2.3. Capacitive Microphone

This is indirectly a modified capacitive pressure transducer in which the vibratory displacements of a diaphragm membrane due to sound pressure variations result in capacitance variations. Sound waves act as pressure waves or fluctuations of air pressure around a normal of one atmosphere. The useful range of this transducer is from 20 Hz to 15 KHz. These are designed to be immune to the effects of variations, if any, in the atmospheric pressure. Atmospheric pressure changes very slowly over time but the changes may be considerably larger than the amplitude of the fluctuating component. The capacitive microphone membrane should be sensitive to the fluctuating components only and special provision must be made to equalize the pressure on either side of membrane so as to prevent it from bursting under variations in atmospheric pressure.

The transducer consists of a very thin metallic membrane kept under radial tension. The capillary leak tube or the hole connecting the pressure chamber below the membrane to the atmosphere is provided to equalize the pressure on either side. This is shown in Fig. 4.10.

The capillary leak and the volume of microphone chamber act as a low pass filter, allowing the slow variations of the atmospheric pressure but preventing the larger sound pressure variations from appearing on the back side of the membrane. By this arrangement, the

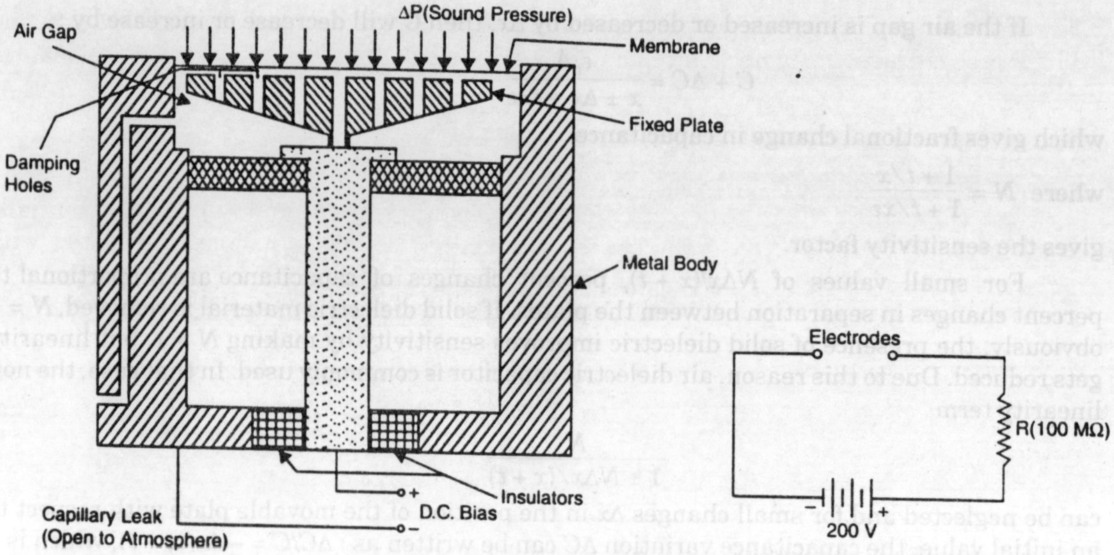

Fig. 4.10. Capacitive microphone.

membrane is subjected to the pressure fluctuations from the front side only. The fixed plate of the capacitor is held rigidly inside the chamber and supported by a column electrode that rests on a metallic body frame. Small openings or damping holes are provided on the plate which allow air to pass freely between the electrodes to the other side of the chamber. The clearance between plates is small and presence of air in between resists the deflection of the membrane. The damping effect of holes also helps in controlling the amplitude of vibration of the membrane at frequencies in the vicinity of its undamped natural frequency. The response of the diaphragm membrane is thus equivalent to that of a spring mass-dashpot system which is flat in the frequency range 10 Hz to 20 KHz. The diameter of the diaphragm membrane varies from 5 mm to 25 mm and the spacing between the plates around 20 μm. The length of the capillary and volume of the microphone chamber should both be small since this considerably reduces the pneumatic time constant. The capacitive variations develop output voltages which vary the high D.C. bias or the polarising voltage

or $$V_t = V_{dc} + V_a$$

V_a **is the voltage signal in audio range impressed upon** V_{dc} **which is normally 200 V.**

Sensitivity is of the order of 1-5 mV per micro bar of sound pressure or about 50 dB.

The output voltage is low and therefore requires to be amplified. The amplifier is normally built inside the microphone casing.

However, the microphone has a low distortion. Resonance frequency is normally 6 KHz. Output impedance is high of the order of 100 mΩ. The total capacitance variation range is about 30 pf.

On flip side, the microphone suffers from the drawbacks given below :

- **It becomes somewhat costly because of the high DC bias.**
- **Cannot withstand excessive heat and moisture.**
- **Separation between plates is rather too small.**

Due to these drawbacks, capacitive microphone is more often used as a standard in sound level meters ; for measuring noise level and for analysis and diagnosis of vibration problems in machinery.

These are also used as a standard for testing and calibration of loud speakers.

4.2.4. Capacitive Hygrometer

This transducer which is used to measure moisture content and humidity, essentially consists of an Aluminium rod acting as an electrode. The rod is oxidised or coated with

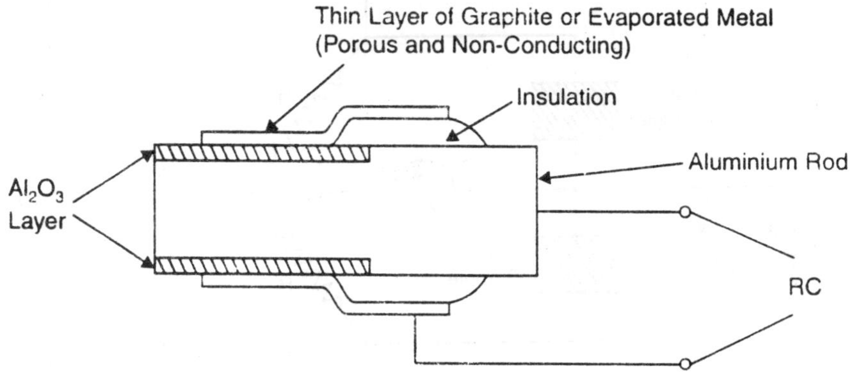

Fig. 4.11. (*a*) Capacitive hygrometer.

Aluminium oxide over a part of its length and is further covered with a thin layer of graphite which acts as a porous layer through which moisture is absorbed and gets to the Aluminium oxide layer. This alters the equivalent capacitance between the outer graphite layer and the Aluminium rod. When moisture has been ingressed and an equilibrium is reached with the humid atmosphere, the resistance and capacitance are both measured. While the resistance falls almost linearly, the capacitance shows an increase but somewhat non linearity. Calibration curves are plotted from which per cent humidity can be computed. Fig. 4.11 shows the arrangement. Calibration curves are also shown.

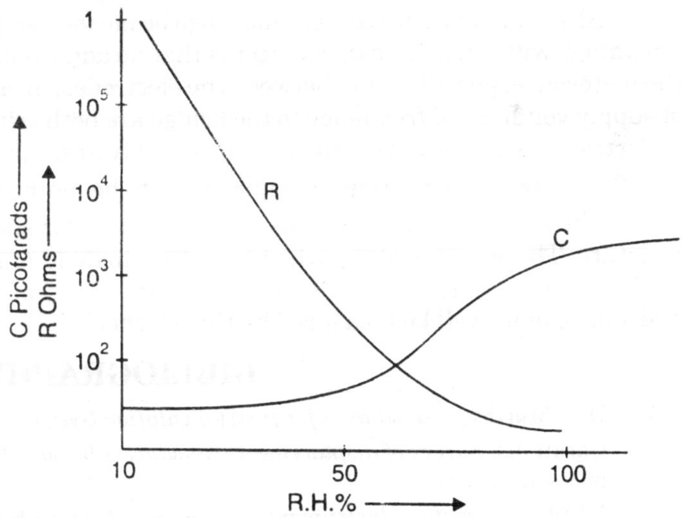

Fig. 4.11. (*b*) Calibration curves variation of R and C with % R.H.

4.2.5. Capacitive Moisture Transducer

Just as capacitive hygrometer measures relative humidity, this particular transducer is designed to sense and measure moisture content in solid and liquid samples. Its operation is based on the fact that the dielectric constant of pure water is about 80 and that of most insulating solids and liquids is less than 10. Therefore, it is possible to measure moisture content in moist solid or its solution in water.

The moisture transducer uses two identical capacitors, one holding the test sample and the other, the same in dry state. The two are connected in two arms of an AC bridge whose opposite arms contain equal resistors R. This is shown in Fig. 4.12. As the capacitance value increases with moisture and its equivalent shunt resistance (representing the dielectric loss component) falls, the arm containing dry sample may be shunted by a capacitor and a resistor, both variable. Their values may be calibrated against the moisture content. Alternatively, the unbalanced voltage may be used as a measure of moisture content.

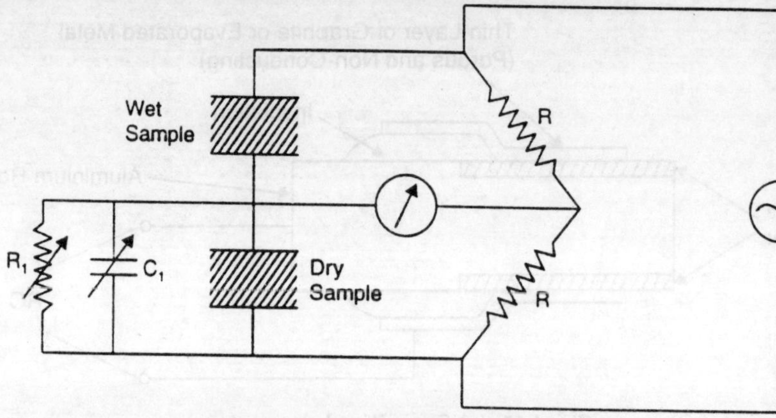

Fig. 4.12. Capacitive moisture transducer.

Although moisture content may also be measured by utilizing variation of resistance, the advantage with capacitive transducer is that no additional efforts are required to compactly fill the material (especially solid) between the electrodes, in order to make a good contact. The choice of supply voltage and frequency to the bridge are both critical as the electrical conductivity and dielectric constant of water change considerably at frequencies above 10 KHz.

Both test and reference samples have to be maintained at the same temperature.

The technique is suitable as well as reliable for such solid materials that readily absorb moisture. Those materials in which moisture instead of getting absorbed deposits only at the surface, are not suitable for this. However, the salinity of water due to addition of contaminants and salts can as well be measured by this method.

BIBLIOGRAPHY

1. DVS Murthy, *Transducers for Instrumentation Systems*, Prentice Hall of India Ltd, 1995.
2. Garatt JD, *Survey of displacement transducers below 50 mm*, J. Phy. E. Scientific Instr., 12(6) pp. 563, July 1979.
3. FJ Oliver, *Practical Instrument Transducers*, Pitman & Sons, London, 1972.
4. HKP Neubert, *Instrument Transducers*, Clarendon Press, oxford, 1975.
5. PCF Wolfendale, *Capacitance displacement transducers with high accuracy and resolution*, J. Sci. Instr. 1, p. 817-18, August 1968.
6. CS Rangan, GR Sarma & VSV Mani, *Instrumentation Devices and Systems*, Tata Mcgraw Hill, 1983.

5

Piezoelectric and Thermoelectric Transducers

The transducers falling into this category are different from those hitherto considered in the sense that these are active transducers or self-generating type, which require no external excitation source. As described in an earlier chapter, these operate on energy conversion principles. These generate an equivalent electrical output signal, *e.g.* from pressure or force to electrical charge and temperature to electrical potential.

5.1. PIEZOELECTRIC PHENOMENON

5.1.1. Transducers Operating on it

Certain solid crystalline substances, *e.g.* quartz, when subjected to mechanical stress, develop an electrical charge, which is proportional to the strain developed. Quartz, which is essentially Silicon dioxide (SiO_2) is available as a natural substance. Its atomic structure is shown in Fig. 5.1. The atoms are arranged as shown, in a hexagonal formation in the plane of paper while *Z*-axis, called optical axis, perpendicular to the *X-Y* plane. For 3 Silicon atoms, 6 Oxygen

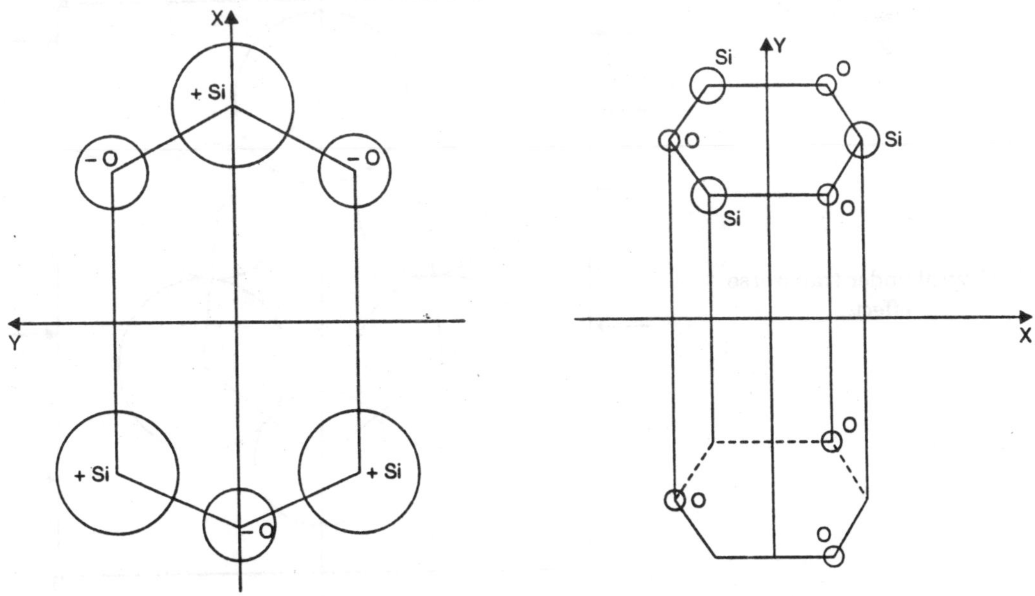

Fig. 5.1. Lattice structure of atoms in a quartz crystal (SiO_2).

atoms are lumped in pairs, thereby forming a hexagonal crystal. X-axis is called electrical axis and Y-axis, the mechanical axis. All the electrical charges are balanced under unstressed conditions but when a force is applied along X-axis, this balance gets distributed and charge develops on two faces A and B. This is called longitudinal effect. A force along Y-axis also distorts the atomic arrangement and charge develops on faces A and B (Fig. 5.2). This is called transverse effect. Because of symmetry along Z-axis or optical axis, no effects are noticed when the force is applied along this.

(a) Crystal under longitudinal effect.

(b) Crystal under transverse effect.

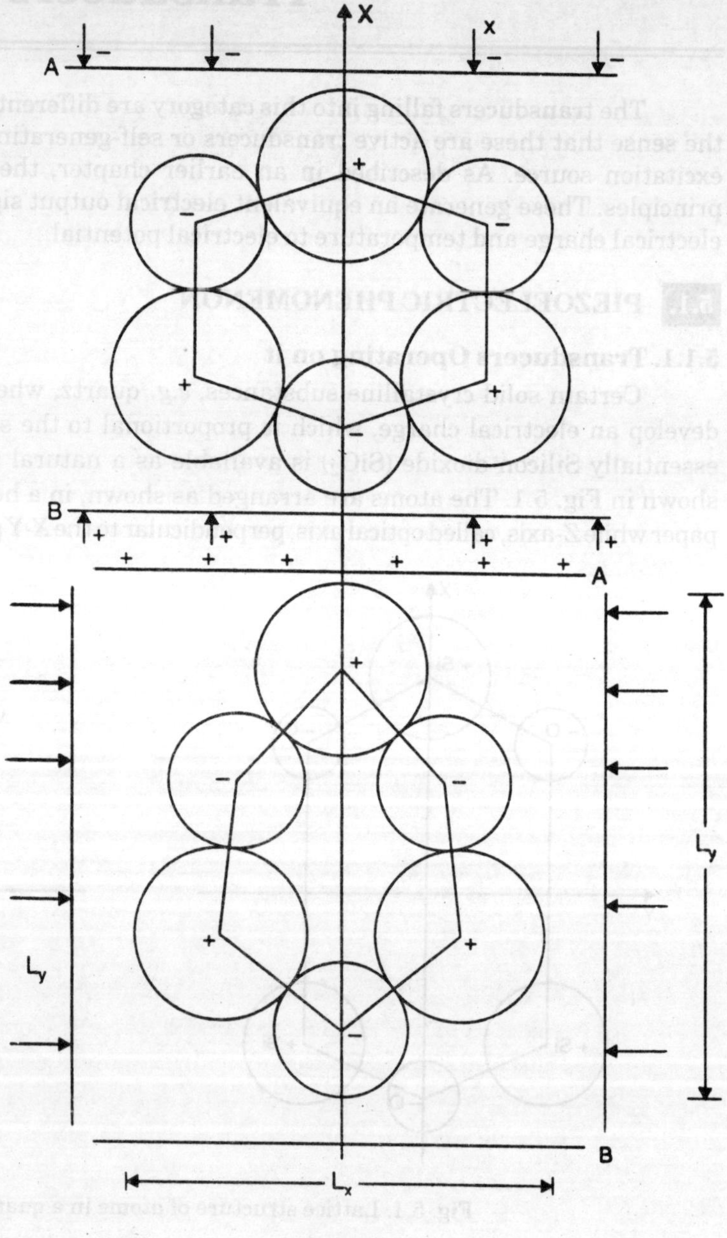

Fig. 5.2

The characteristic features of the longitudinal effect are that the charge generated is independent of the area of crystal and its thickness in X direction. The charge developed on a given area of crystal face is proportional to the area affected by the pressure or proportional to the total force applied normal to the surface.

However, when the force is applied in the transverse or Y direction the charge developed on faces A and B depends on the length L_x in X direction and length L_y in Y direction.

Quartz being an insulator, metallic electrodes or gold coating used as electrode are required to serve as a link between the charge accumulated and the measuring circuits.

If l and b are the length and breadth of the coated surface, its area $A = lb$.

Therefore, the transducer capacitance

$$C_x = \varepsilon_o \varepsilon_r A/t$$

where t = crystal thickness and

ε_o, ε_r the permittivities of air and the solid dielectric and the voltage across the electrodes

$$V_x = Q_x/C_x = d_{11}F_x/C_x = d_{11}\sigma_{xx}t/\varepsilon_o\varepsilon_r = g_{11}\sigma_{xx}t \quad ...(5.1)$$

where d denotes the coefficient of charge sensitivity.

The first subscript of d_{11} signifies the direction of electrical effect and the second subscript, that of mechanical effect.

σ_{xx} = Axial stress along X-axis.

F_x = The force applied along X direction.

g_{11} = The voltage sensitivity expressed as $V_x/\sigma_{xx}t$ which denotes the electric field produced unit stress.

If force F_y is applied along Y-axis, the charge and electric field developed in X direction are given by :

$$Q_x = d_{12}F_y = d_{12}A\sigma_{yy} \quad ...(5.2)$$

and since

$$C_x = \varepsilon_o \varepsilon_r A/t$$

$$V_x = \frac{Q_x}{C_x} = \frac{d_{12}A\sigma_{yy}t}{\varepsilon_o\varepsilon_r A} = \frac{d_{12}\sigma_{yy}t}{\varepsilon_o\varepsilon_r} \quad ...(5.3)$$

and

$$g_{12} = \frac{V_x/t}{\sigma_{yy}} = \frac{d_{12}}{\varepsilon_o\varepsilon_r} \quad ...(5.4)$$

and d_{12} = Charge sensitivity when stress is developed by the force applied in Y direction

σ_{yy} = Axial stress developed along Y-axis

g_{12} = Voltage sensitivity given by electric field per unit stress.

5.1.2. Piezoelectric Materials

Piezoelectric materials fall into two groups : natural and synthetic. While Quartz, Rochelle salt and tourmaline are naturally occurring, Ammonium dihydrogen phosphate (ADP), Lithium sulphate and Potassium tartarate are among the synthetic.

Quartz is natural and most stable. It is also grown artificially and is preferred since it is purer than the naturally occurring one. Its charge sensitivity (d coefficient) and permittivity ε_r remain unaffected by temperature. Only its volume resistivity falls sharply at temperatures above 100°C. Tourmaline is the only material exhibiting large sensitivity in the volume expansion mode related to Z-axis. Its useful temperature range is upto 1000°C.

Rochelle salt is produced on an industrial scale for the manufacture of gramophone pick-ups and crystal microphones. Its permittivity ε_r is the highest among those falling in the natural group.

ADP crystals possess the lowest resistivity, which is also temperature dependent. Hence these are used for making acceleration and pressure transducers with proper temperature compensation.

To choose between Quartz and Rochelle salt, the former, though stable, offers smaller output. The latter, though giving the highest output, has its permittivity ε_r humidity sensitive. Barium titanate is yet another piezoelectric material that can be shaped into a variety of forms.

5.1.3. Piezoelectric Pressure/Force Transducer

A piezoelectric element can convert mechanical motion into electrical signals and acts as a charge generator as well as a capacitor. Mechanical deformation generates a charge which appears as a voltage across the electrodes, $E = Q/C$.

The piezoelectric effect is direction sensitive. Tensile and compressive forces produce voltages of opposite polarity. The magnitude and polarity of the induced surface charges are proportional to the magnitude and direction of the applied force

i.e. $Q = dF$ Coloumbs

where d = Charge sensitivity of crystal in Coloumbs/Newton which is constant for a given crystal and F is the applied force in Newtons.

The force causes a small change δt of the crystal.

$$F = \frac{AY}{t}\,\delta t \text{ Newtons} \quad ...(5.5)$$

where Y = Young's modulus in N/m^2

A = Area of the crystal in m^2

T = Thickness of crystal in m.

But $$Y = \frac{F/A}{\delta t/t}\ \text{N/m}^2 = \frac{Ft}{\delta t A} \quad \text{or} \quad F = \frac{AY}{t}\,\delta t$$

Therefore $$Q = \frac{dAY}{t}\,\delta t \quad ...(5.6)$$

This charge at the electrodes rises to a voltage output V_o

Therefore $V_o = Q/C_o$ but since $C_o = \varepsilon_o \varepsilon_r A/t$

$$V_o = \frac{Qt}{\varepsilon_o \varepsilon_r A} = \frac{dFt}{\varepsilon_o \varepsilon_r A} \quad ...(5.7)$$

Substituting $dAY\,\delta t/t$ for dF

$$V_o = \frac{dY\delta t}{\varepsilon_o \varepsilon_r} = \frac{d\delta t Y}{\varepsilon_o \varepsilon_r} = \frac{d\delta t}{\varepsilon_o \varepsilon_r} \cdot \frac{Ft}{A\delta t}$$

or $$V_o = \frac{d}{\varepsilon_o \varepsilon_r} \cdot \frac{Ft}{A} = \frac{d}{\varepsilon_o \varepsilon_r} \cdot Pt = KPt \quad ...(5.8)$$

where $K = d/\varepsilon_o \varepsilon_r$ is the voltage sensitivity of the piezoelectric crystal.

or $K = V_o/tP = (V_o/t)/P$ = Electric field/Stress.

5.1.4. Pressure Transducer Producing Frequency Change

When a piezoelectric quartz crystal is used for measurement of force, it gets strained and its thickness gets slightly reduced. If such a crystal is used as a resonator for a tuned oscillator, a variation in frequency is observed. When the pressure applied changes by one atmosphere, a variation of 2 Hz is noticeable in the oscillator frequency. The temperature of the quartz crystal is maintained constant and the change in frequency is measured precisely by an arrangement described later.

Pressure ranging between 0-700 atmosphere can be measured by the method with a resolution of 1 in 10^6 for a one second counting period. The output signals are converted from analog form to digital form (frequency or time domain), somewhat indirectly.

5.1.5. Quartz Temperature Transducer

A piezoelectric quartz crystal can be made to measure temperature with good accuracy and high stability. The temperature frequency relationship is highly linear with excellent repeatability in the range 193-573K. The linearity is in fact, better than all other methods of temperature measurement. The frequency *vs.* temperature relationship takes the form :

$$f_T = f_o(1 + aT + bT^2 + cT^3 + ...) \quad ...(5.9)$$

where, f_o is the oscillator's fundamental frequency at 0°C and f_T, that at temperature T °C. If the quartz crystal is cut in a specific direction the coefficients b, c...... become negligible and the relationship becomes quite linear and simpler of the form :

$$f_T = f_o(1 + aT)$$

where the coefficient $a = 35.4 \times 10^{-6}$/°C

f_o depends on the crystal thickness, and can be adjusted to a value that yields a frequency variation of 1 KHz for a 1°C change of temperature. This holds good when the resonator frequency f_o is 28.2 MHz. This is called the reference frequency and at this particular crystallographic orientation, the lowest frequency shear mode of operation is ensured.

Within the range – 40 to + 250°C, the device is linear within ± 0.05%. The sensing crystal of quartz is hermatically sealed in a cylindrical copper case in a helium (inert gas) atmosphere. The case itself is further enclosed in a stainless steel tube as in thermocouples. With a suitable design of the resonator and a closely coupled probe, a response time as low as 1 second is obtained.

As mentioned above, a 1°C change in temperature causes a frequency change of $28.2 \times 10^6 \times 35.4 \times 10^{-6} = 1000$ Hz. Hence for counting for 0.1 second, on a five-digit counter, the temperature will be correctly indicated, with a resolution of 0.01°C.

Difficulty of achieving a high resolution of this order is obviated by having another quartz oscillator operating at a fixed but known temperature and comparing the two frequencies by using a frequency mixer. The circuit arrangement is shown in Fig. 5.3, through a block diagram.

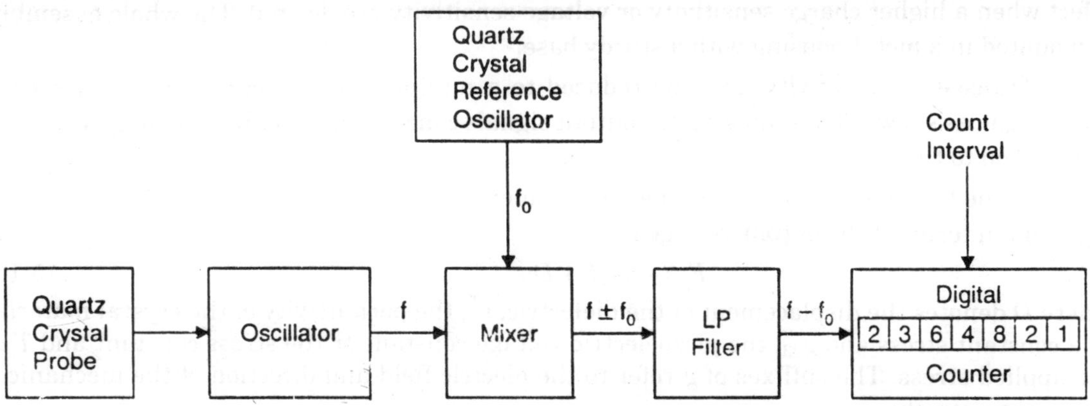

Fig. 5.3. Block diagram of a quartz crystal temperature transducer.

5.1.6. Biomorphs, Multimorphs or Piezopiles

When more than one piezoelectric elements are put in random, these form biomorphs (two elements), Multimorphs (more than two elements) or piezopiles. Normally, two or four elements are joined mechanically in series but electrically, these are in parallel. This increases their net capacitance two or four fold and accordingly, their charge sensitivity also increases by the same factor.

The series connection (electrically) may result in a higher voltage sensitivity, provided the resulting reduced capacitance is a satisfactory match to the signal processing circuits. Electrical parallel connection, of course results in an increased charge sensitivity.

These are illustrated by diagrams in Fig. 5.4.

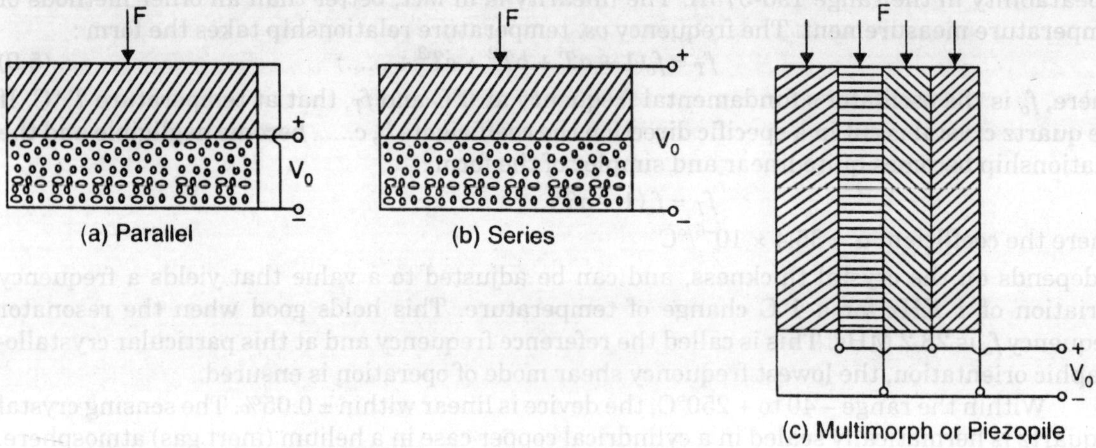

Fig. 5.4. Biomorphs and multimorphs.

5.1.7. Piezoelectric Accelerometer

This is essentially a piezoelectric force transducer with a proof mass for developing force under the acceleration input. Being an active transducer, it is a self-generating one and thus converts the dynamic force or acceleration into electrical energy.

The accelerometer is fabricated with a mass mounted in direct contact with the crystal. It is preloaded or pre stressed by screwing down the cap on the hemispherical spring, as shown in Fig. 5.5. In place of a single crystal, even a piezopile may also be used to have a multiplying effect when a higher charge sensitivity or voltage sensitivity are desired. The whole assembly is mounted in a metal housing with a sturdy base.

Transverse sensitivity must be reduced to a minimum by strong guide springs or a diaphragm, as shown. Thus, making the output signals cancel each other by providing a specific orientation to their *Y*-axis.

In the frequency range where the accelerometer sensitivity is constant, the piezoelectric equation in terms of the output voltage is

$$E = -g_{33}T + D/\varepsilon^T \qquad \text{...(5.10)}$$

where D denotes the displacement of the dielectric, ε^T, the permittivity of the crystal material at a constant stress and g_{33}, the piezoelectric voltage constant or the stress constant and T is the applied stress. The suffixes of g refer to the electric field and direction of the mechanical stress respectively.

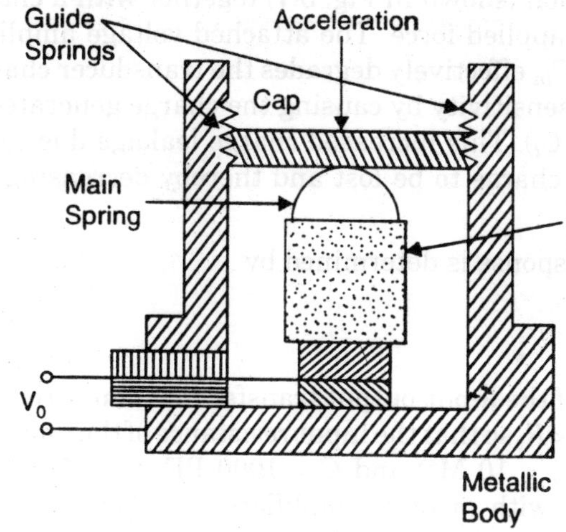

Fig. 5.5. Piezoelectric accelerometer.

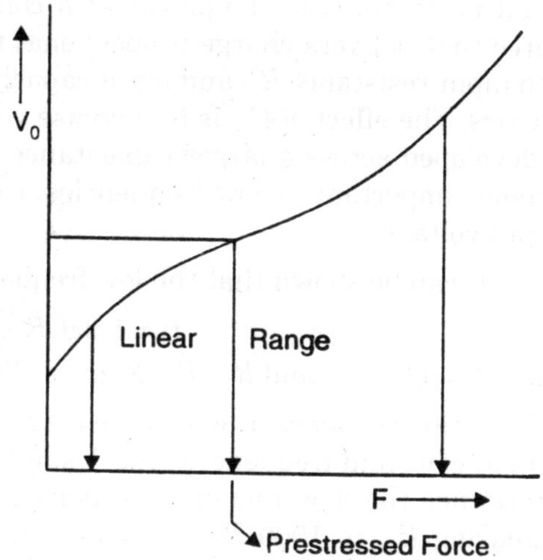

Fig. 5.6. Frequency response.

where the proof mass (or the seismic mass) m is many times greater than the mass of the piezoelectric crystal, the mechanical stress T is given by

$$T(t) = a(t)\ .\ m/a \text{ (in compression)}$$

where $a(t)$ is the acceleration and A, the cross-sectional area of the piezoelectric crystal.

The piezoelectric accelerometer gives a higher sensitivity and bandwidth but has a slight departure from linearity and stability. In case, a piezopile is employed in place of a single crystal, the self mass (distributed) of the biomorph itself acts as a proof mass. This has been shown in Fig. 5.6.

5.1.8. Low Frequency Response of Piezoelectric Signals

The piezoelectric materials have a high but finite resistivity. As a result, the charge generated by a change in applied force slowly leaks away, causing the potential difference to reduce eventually to zero. Quartz, for example, has a resistivity higher than 10^{12} ohm-metre. Therefore, effectively, pizoelectric transducers lack a DC response and only dynamic forces or those varying in time are capable of yielding an output. At medium and low frequencies, the

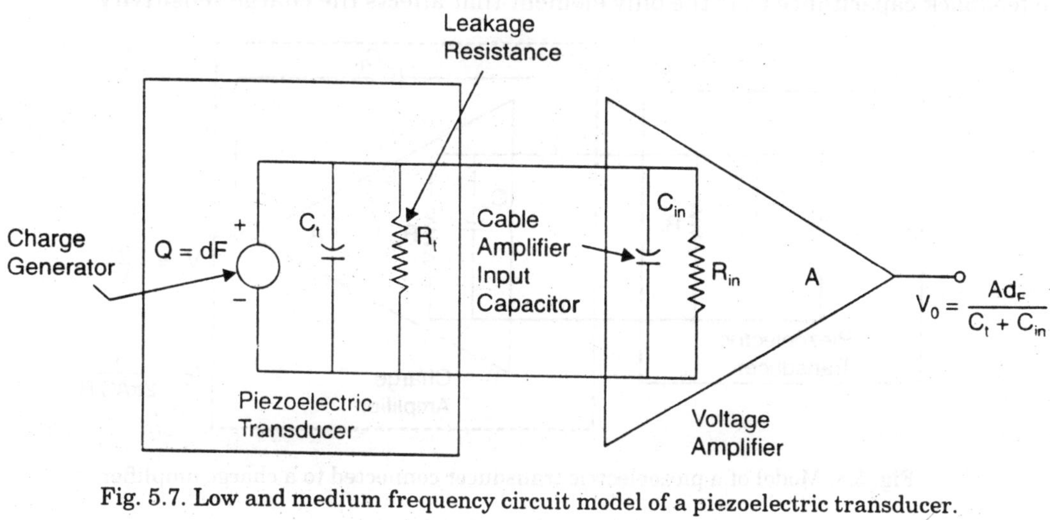

Fig. 5.7. Low and medium frequency circuit model of a piezoelectric transducer.

circuit model consists of a parallel CR combination (shown in Fig. 5.7) together with a charge source that delivers charge proportional to the applied force. The attached voltage amplifier with input resistance R_{in} and input capacitance C_{in} effectively degrades the transducer characteristics. The effect of C_{in} is to increase voltage sensitivity by causing the charge generated to be developed across a larger capacitance ($C_{in} + C_t$). This increased charge leakage due to R_{in} becomes important at low frequencies, causing charge to be lost and thereby decreasing the output voltage.

It can be shown that the low frequency response is determined by

$$f_c = 1/2\pi CR$$

where $C = C_{in} + C_t$ and $R = R_{in} R_t/R_{in} + R_t$

and f_c is the frequency at which the magnitude of the input-output transfer function is reduced to 0.707 of its mid-frequency value. Usually, $R_{in} << R_t$ so that the input resistance of the amplifier determines the low frequency response. For R_{in} = 10 MΩ and C = 1000 PF, f_c = 17 Hz. By increasing R_{in} to 1000 MΩ (readily achievable with modern amplifiers), the low frequency response is reduced to 0.17 Hz.

5.1.9. Use of a Charge Amplifier

A voltage amplifier shown above is useful for amplifying the output voltage signals provided it has a large input resistance and small input capacitance in comparison with the charged capacitor. However, when very slowly varying stresses are required to be measured, an amplifier suitably designed to develop output voltage signals truly proportional to the charge as also to the stress, is required. In other words, the amplifier bandwidth should extend to include very low frequencies, say 0.01 Hz.

A charge amplifier that converts the transducer charge into a proportional output voltage, offers several advantages over the voltage amplifier. Its very large input capacitance makes the sensitivity of such a system essentially independent of both the transducer and the connecting cables capacitance. Also, since the voltage developed across the amplifier input is greatly reduced, the effect of the amplifier input resistance and the transducer loss resistance on charge leakage becomes very small. Fig. 5.8 illustrates how the charge amplifier is connected to the piezoelectric transducer. It is to be noted that if the gain of the operational amplifier is large, the feedback capacitance C_f is the only element that affects the charge sensitivity.

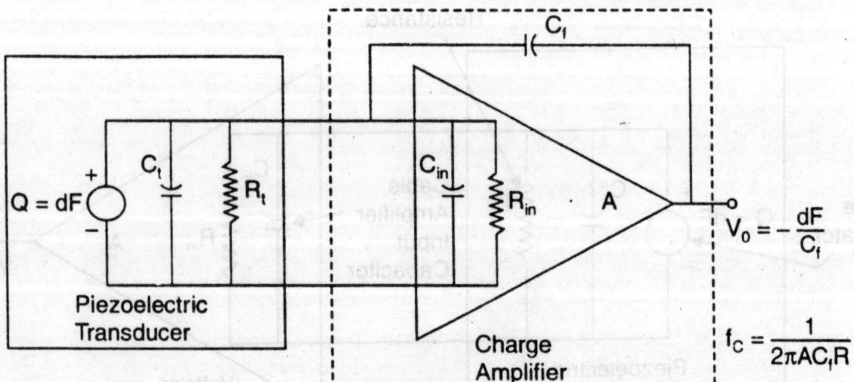

Fig. 5.8. Model of a piezoelectric transducer connected to a charge amplifier.

5.1.10. Piezoelectric Pressure Transducer for Under Water Pressure Measurement

The special category of piezoelectric transducer for measuring under water and air blast pressures should ideally be a non-directional one. A hollow spherical shape is therefore the best choice. Its dimensions should be small compared with the wavelength of air blast or water. In most cases a small hollow cylinder is used instead. Its outer and inner surfaces are metallized and are used as electrodes. The pressure is applied to the outer surface. The piezoelectric crystal (usually barium titanate) of which the hollow cylinder is made, responds to pressure P_e in all three modes, *viz.* thickness mode, length mode and radial mode. For a thin walled hollow cylinder, open-circuit voltage generated by radial and tangential stress is given by :

$$V_o = P_e \left[K_{\text{radial}} \cdot b \frac{a-b}{a+b} - K_{\text{tangential}} \cdot b \right] \quad ...(5.11)$$

where a and b are the inner and outer radii of the hollow cylinder. Since K_{radial} and $K_{\text{tangential}}$ are of opposite signs, $(a - b)$ is negative. Therefore V_o becomes zero at some critical value of the ratio b/a. A proper choice of the ratio wall thickness/outer radius or b/a is necessary. For low values of applied pressure P_e, the value of this ratio is kept below 0.20 and for higher values of P_e, it is to be well above 0.35, thereby avoiding the critical region at which V_o becomes zero.

The transducer is shown in Fig. 5.9.

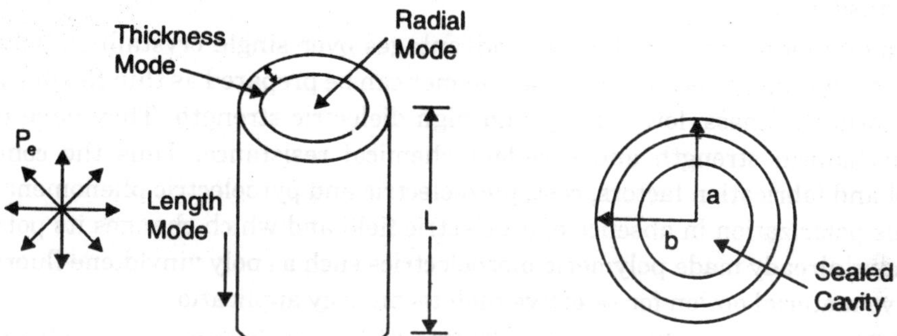

Fig. 5.9. Piezoelectric transducer for under water and air blast pressures.

5.1.11. Piezoelectric Polymers

The piezoelectricity in polymers has been found in a number of oriented synthetic as well as biological polymers and is ascribed to piezoelectric properties of crystallites embedded in amorphous regions of polymers.

Polymeric piezoelectric materials offer many salient features such as availability in the form of thin, uniform and flexible films of large area, low density and high dielectric strength. These have also low mechanical and acoustic impedance almost of the same magnitude to that of water and consequently of human body. They also have the advantage of ease of fabrication at low cost. Though piezoelectricity was discovered in both biological and synthetic polymers in the late 1950s and early 1960s, it was discovery of piezoelectricity in poly vinylidiene fluoride (PVDF) by kawai during the late 1960s, which intensified the interest in polymeric piezoelectrics. The presence of large dipole moment in PVDF plays an important role in determining the pronounced piezoelectric effect in this polymer. A very high piezoelectricity has been reported in the amorphous copolymers of vinylidene cynide and vinyl acetate (PCVDCN-VAC) which may originate from C ≡ N dipole moments.

Very recently, the existence of piezoelectricity in liquid crystals has been reported. (around a decade ago).

The theory of piezoelectricity in polymers is centered around PVDF. A change in polarisation (dipole moment per unit volume) can occur through changes in either the moment or the volume, with aligned dipoles present, can yield large polarisation changes through the changes in volume. However, most of the proposed mechanisms for piezoelectric effect in PVDF focus on the changes in the moment. Various other models have recently indicated that the largest contribution to the activity of PVDF arises from bulk dimensional changes rather than from changes in molecular dipole moments. The advent of (PCVDCN-VAC) having the same order of piezoelectricity which is of amorphous nature has made all the theories more speculative than real.

The piezoelectric materials exist in various forms and kinds, *i.e.* crystalline, polycrystalline ceramics, polymeric ceramics, polymeric composites, glass and thin films on substrate.

Quartz is an example of crystalline piezoelectric. Among piezoelectric ceramics, lead zirconate titanate (PZT) is prominent. A large number of ceramic composites have been formed by dispersing active PZT materials in plastics or elastomers. The PZT and other ceramic piezoelectrics can also be deposited by sputtering on metal and non-metallic substrate to form thin film transducers.

Polymeric piezoelectrics offer real advantages over single crystalline, polycrystalline, glassy and ceramic piezoelectrics because polymer can be prepared as thin flexible film of large area of uniform thickness, low density and high dielectric strength. They have high processability, mechanical strength and excellent chemical resistance. Thus the combination of mechanical and fabrication factors, cost, piezoelectric and pyroelectric phenomena (exhibiting spontaneous polarization in absence of an electric field and which changes its polarization on heating), have already made polymeric piezoelectrics such as poly vinyldiene fluoride (PVDF), an attractive material option for selective high technology applications.

PVDF is a semi crystalline polymer (50%) which exists in various crystalline phases and forms crystals typically 10 nm thick by 100 nm long dispersed in an amorphous phase which has super cooled liquid properties and a glass transition at 40°C. The referred form for piezoelectricity is the β form (which has planar zig zag conformation and ordered C-F bonds giving large unit dipole and high dielectric constant). The properties of β form may be increased by special treatment during extrusion process. PVDF films are uniaxially or biaxially stretched or rolled for converting one form to another and to also improve mechanical and electrical properties.

In order to impart piezoelectric properties to PVDF, the material is initially stretched and annealed and then poled either thermally or by corona discharge method. In case of thermal poling, electrodes are deposited onto PVDF and it is subjected to a field of 0.4-1.40 MV/cm at temperature around 90-100°C. The dipolar alignment occurs by reorientation of randomly oriented crystalline regions in the direction of the field. The orientation is caused by chain orientation of the crystallites which are interconnected by long molecular chains and thus hindered in their rotation.

The parallel orientation of these dipoles leads to a residual polarization in the film which is the basis of several mechanisms proposed for existence of piezo and pyroelectricity in the PVDF.

Applications

A large number of applications have been suggested for polymeric piezoelectrics, particularly for PVDF. These applications are : Sonar hydrophone, ultrasonic transducers, audio frequency transducers, pyroelectric sensors and electro-chemical devices. PVDF, with its low acoustic impedance and large piezoelectric coefficient is ideally suited as a transducer for broadband under water receivers in hydrophones. PVDF has very wide band characteristics and has been used as transducer at frequencies from DC upto 1 GHz. The transducers are prepared from 25 μm thick films with areas upto 100 cm^2, which are freely mounted or applied with an adhesive to a solid or flexible back plate. The design of ultrasonic transducers for use in pulse echo systems is another area of application in which PVDF transducers show definite advantages over ceramic transducers. In pulse echo systems, the transducer sends upto an ultrasonic pulse in the frequency range of 1-10 MHz and returned pulse is analysed to provide the required information. These transducers are used in non-destructive testing and medical purposes. As a transmitter of acoustic power, PVDF transducer is quite poor but its increased sensitivity on reception is particularly valuable in medical applications such as foetal examination where it is essential to minimize the level of acoustic power transmitted into patient body. Due to very large bandwidth, very sharp ultrasonic pulse is received by the transducer giving better resolution. An array of transducers may by prepared on a single PVDF sheet.

Monolithic Silicon-PVDF devices have been prepared where a sheet of PVDF is bonded to a Silicon wafer containing an array of metal oxide semiconductor field effect transistor (MOSFET) amplifiers. The device is so arranged that where an acoustic wave is detected, the electrical signal resulting from the piezoelectric action in the PVDF appears directly on the gate of a MOS transistor. Such device is called POSFET for use in medical imaging. Using this device, a linear array of 34 elements has been manufactured with a 6 MHz bandwidth and a dynamic range of over 70 dB. This range is expected to be further broadened with continued R and D efforts in this direction.

However, in view of somewhat low pyroelectric coefficient, PVDF is being extensively used where speed and resolution are more important rather than sensitivity. For example, in pyroelectric vidicon where one can obtain very high resolution pictures by using PVDF since its low conductivity can reduce thermal spreading.

PVDF has been suggested as a material for use as biomorph structure for display devices. These display devices use PVDF cantilevers which deflect when voltage is applied to the film. This deflection is used to control transmission, reflection or absorption of light. Deformable mirrors and coin sensors have also been suggested as potential applications.

5.2. THERMOELECTRIC TRANSDUCERS (THERMOCOUPLES)

When a closed circuit is formed by joining conducting leads of two dissimilar metals and the two junctions are kept at different temperatures, electrical current starts flowing. The concentration of valence electrons in each lead is proportional to the temperature difference and at the junction or the point of contact, these valence electrons diffuse through the boundary layer between the two leads, making one lead positive and the other negative. Thus the emf generated is proportional to the temperature difference between the two junctions. This phenomenon is known as Seebeck effect. The seebeck coefficient or sensitivity S of a thermal element is given by

$$S = e/t = a + bt + ct^2$$

(S is also sometimes called thermoelectric power)

where e is the emf generated, t is the change in temperature and a, b, c are constants determined by measuring the value of e at three standard reference temperatures.

The magnitude of S depends upon the chemical composition and the physical treatment of the materials used. The Table 5.1 below gives its nominal values for various materials at 273.15°K when compared with Platinum :

Table 5.1

Material	S, μV/°C
Bismuth	– 72
Constantan	– 35
Alumel (Nickel)	– 15
Platinum	0
Aluminium	3.5
Rhodium	6.0
Copper	6.5
Gold	6.5
Iron	18.5
Chromel	25.0
Silicon	440

In practice, the temperature measurement is carried out in such a way that one junction is kept at a constant temperature (usually 273.15°K) and the other is exposed to the medium whose temperature is to be measured.

The properties of a few thermocouples are given in Table 5.2. The ultimate limits are set to the temperatures at which oxidation or reduction of materials starts.

Table 5.2. Properties of Thermocouples

Type	*Composition*	*Sensitivity* μV/°C	*Range* degrees C	*Accuracy*	*Observations and Remarks*
E	+ leg : Chromel 90% Ni, 10% Cr		– 18 to 315	± 1.8°	High sensitivity
	– leg : Constantan 57% Cr, 43% Ni	40-55	315 to 870	± 0.5%	Deteriorates slowly in reducing environment
K	+ leg : Chromel 90% Ni, 10% Cr		– 18 to 276	± 2.2°	Long life Low thermal conductivity
	leg : Alumel 94% Ni, 3% Mn, 2% Al, 1% Sr	40-55	276 to 1000	± 0.5%	Most popular, wide applications Good linearity
J	+ leg Iron Fe 100% – leg Constantan 57%, Cr 43% Ni	45-57	– 18 to 276 276 to 560	-do-	Inexpensive Mechanically strong Deteriorates rapidly above 560°, Rusts easily

T	+ leg Copper Cu 100% – leg Constantan 57%, Cr 43% Ni	15-60	– 200 to 93 93 to 350	± 0.8° ± 0.5%	High reliability, especially to low temperatures below 0°C, Suited to reducing environments
R	+ leg Platinum Rhodium Pt 87%, Rh 13% – leg Pt 100%	5-12	0-1000	± 0.25%	Susceptible to Corrosion Good linearity
S	+ leg : Plantinum-Rhodium Pt 90%, Rh 10% – leg : Platinum Pt 100%	5-12	0-1500	± 0.25%	Low sensitivity, high stability Free from parasite emfs can be used in oxidising environment
B	+ leg : Indium-Rhodium 40% In, 60% Rh – leg : Indium In 100%	5	1200-2000	± 1°	-do-

Fig. 5.10 gives representative values for thermo emfs as a function of temperature difference for the above popular types of thermocouple, over their respective ranges.

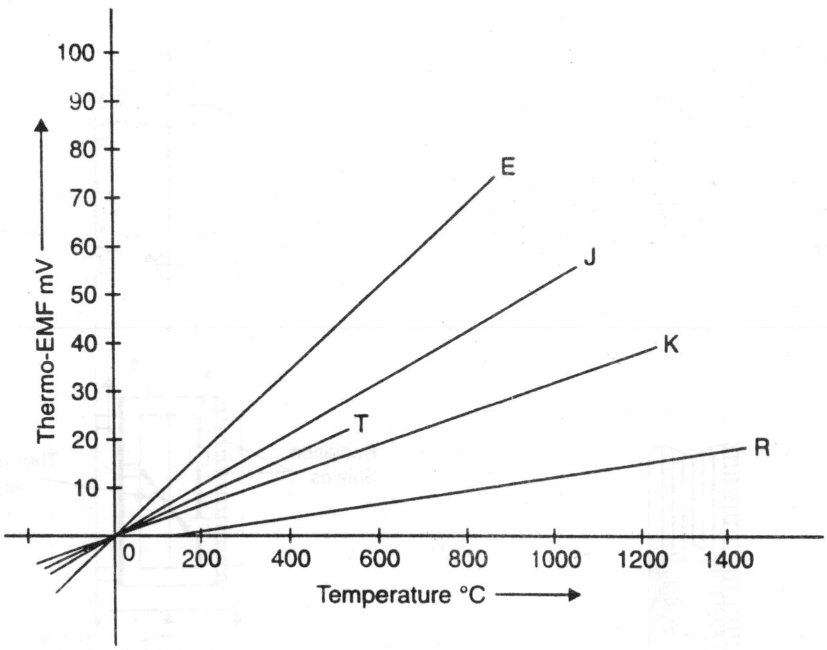

Fig. 5.10. Thermo-emf *Vs* temperature for different types of thermocouples.

5.2.1. Constructional Features

Depending upon the response time, insensitivity to vibration and corrosion resistance, different construction techniques are adopted for the thermocouples. Fig. 5.11 illustrates some techniques. At (*a*) is shown an open twisted type where the hot junction is made by twisting the two wires, at equal angles. Here the hot junction is directly exposed to the medium. Obviously,

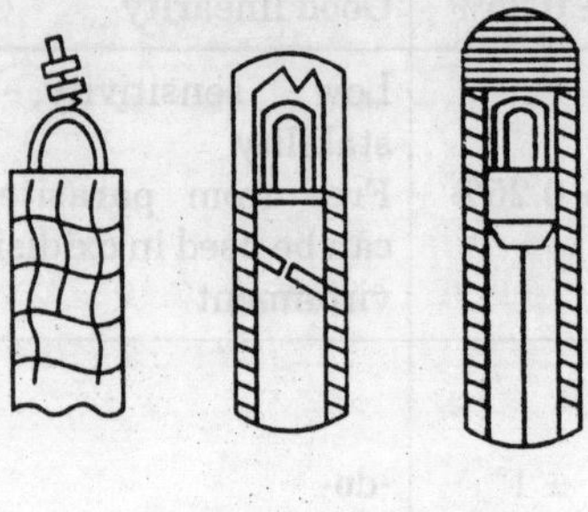

Fig. 5.11. Various types of construction for thermocouple probes.

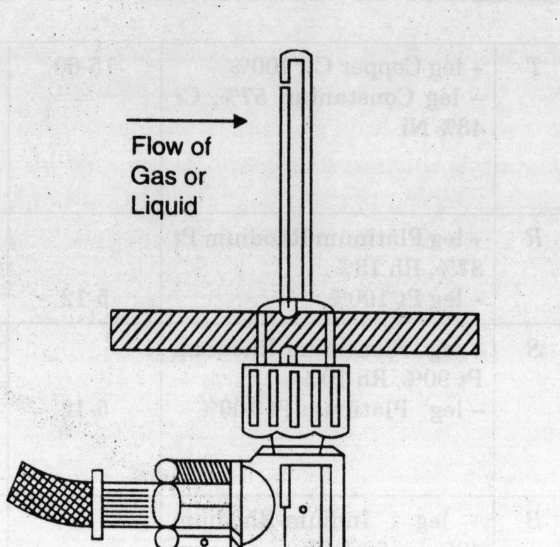

Fig. 5.12. Thermocouple probe for liquid/gas flow.

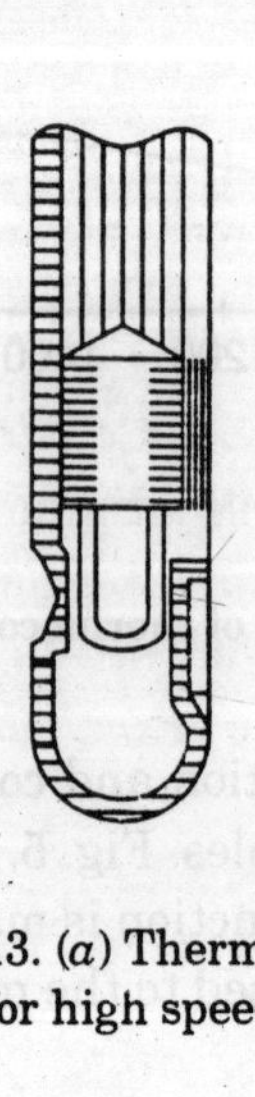

Fig. 5.13. (*a*) Thermocouple probe for high speed gases.

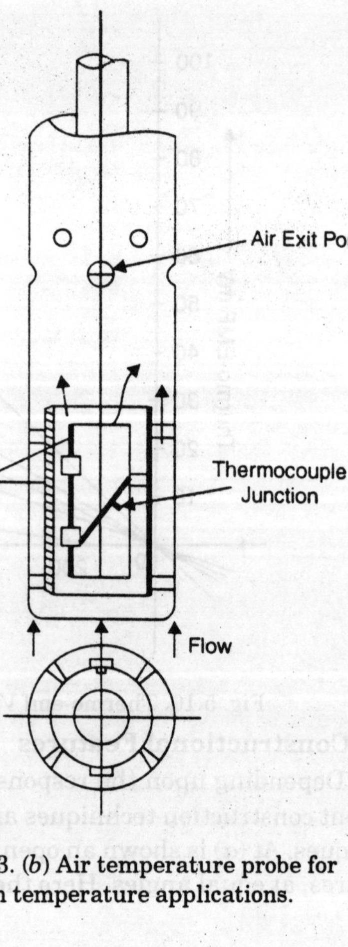

Fig. 5.13. (*b*) Air temperature probe for high temperature applications.

the response time is small. Instead of twisting, welding (by non-oxidising, neutral flame) or silver soldering are also adopted. Fig. 5.11 (*b*) shows the element wires welded and grounded to the enclosing bulb at the contact point and at 5.11 (*c*), it is the closely ungrounded element. Response time for configuration at (*c*) is lower than at (*b*). When the thermocouple is used for measuring temperature of a fluid in motion, the open element is usually protected with a shield.

When a thermocouple is enclosed in housing assembly, it is usually termed thermocouple probe. These probes have lead wires without terminals or connectors with pins made out of the same thermoelement material. Probes for measurements in liquids and gases are usually cylindrical in shape and the threaded fitting is often manufactured with a clamp, so as to make the immersion depth of the probe adjustable. This is shown at Fig. 5.12. For measuring temperatures in rapidly moving gas streams, special types of probes are used, as shown at Fig. 5.13 (*a*), best suited for measuring temperature in rapidly emerging exhaust gases. The total air temperature probes with concentric radiation shields (for reducing large radiation errors), suitable for measuring outside air temperatures in a supersonic aircraft, is shown at Fig. 5.13 (*b*).

5.2.2. Measuring Circuits used with Thermocouples

Referring to Fig. 5.14 (*a*), where two different metals *A* and *B* are joined to form the thermocouple junctions T_1 and T_2. If these junctions are at the same temperatures, no thermo-emf is produced in the circuit. However, if the junction T_2 is maintained at 0 °C or 273.15 °K and the temperature of the junction T_1 is raised, the thermo emf generated, which is independent of the temperature gradient of the lead wires, is a function of the temperature difference

$$\Delta T = T_1 - T_2.$$

If one of the conductors is broken and a millivoltmeter is inserted [Fig. 5.14 (*b*)], with the leads made of a third metal say, Copper and Leads are kept at the same temperature T_3 there will be no effect on the emf generated although two new but differentially arranged thermo-couples are introduced. However, in practice, when the temperature of the junction T_1 is desired, instead of opening one lead and using two differentially connected thermocouples, both the leads are opened and the millivoltmeter inserted as shown in Fig. 5.14 (*c*).

If the leads of the thermocouple are short and the temperature T_1 is relatively high, it would be difficult to maintain a constant temperature at the measuring junction because there may be large and varying temperature gradients between the junction and the indicator. In such cases, separate reference junctions T_{12} and T_{13} located at a more favourable point away from T_2, with extension leads, is adopted as shown at Fig. 5.14 (*d*).

When Chromel Alumel thermocouples are used, there will be a fairly high drop in voltage generated if long leads are employed. Leads extensions with Copper-Constantan wires can be used to minimise the effect of lead resistance. In case of Platinum/Platinum-Rhodium thermo-couple, Copper and Copper-Nickel alloys are best suited extension lead materials since these are economic also. The thermo emf of the leads remains small well within a limited range of temperature. The effect of lead resistance becomes negligible if the measuring instrument with a high input impedance is used.

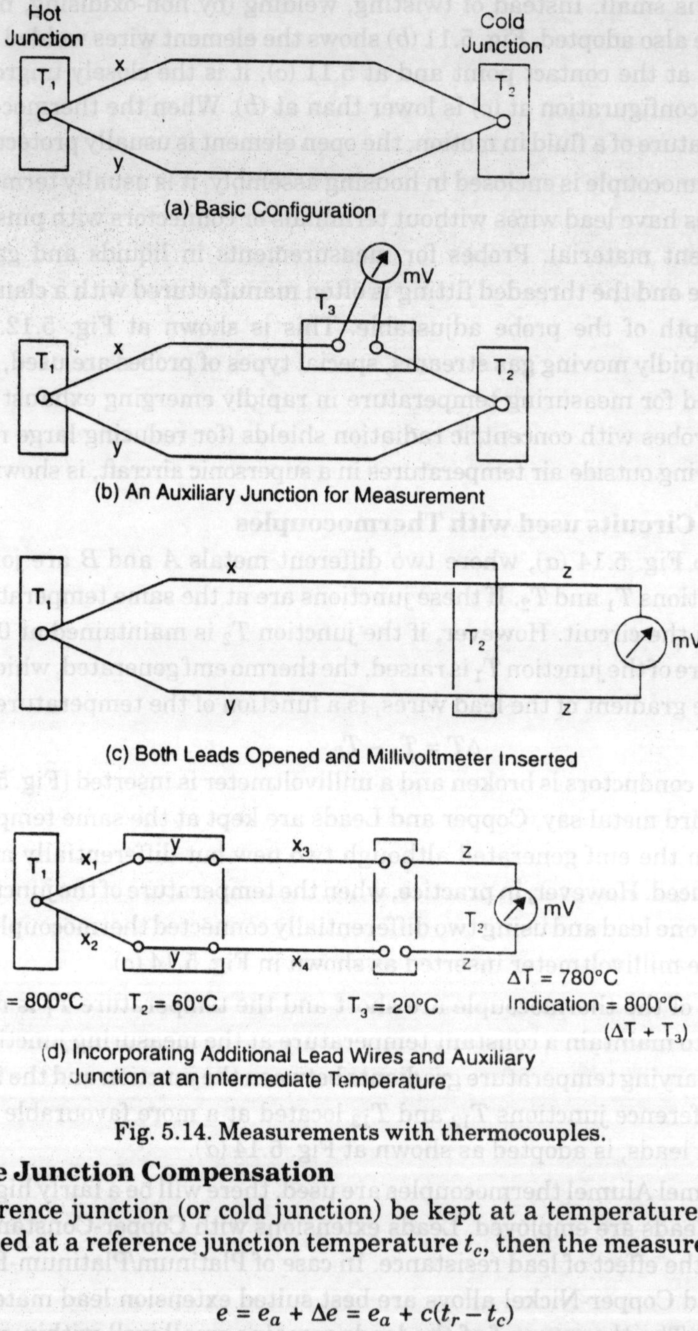

(a) Basic Configuration

(b) An Auxiliary Junction for Measurement

(c) Both Leads Opened and Millivoltmeter Inserted

(d) Incorporating Additional Lead Wires and Auxiliary Junction at an Intermediate Temperature

Fig. 5.14. Measurements with thermocouples.

5.2.3. Reference Junction Compensation

If the reference junction (or cold junction) be kept at a temperature t_r but the thermocouple is calibrated at a reference junction temperature t_c, then the measured thermo emf may be corrected as :

$$e = e_a + \Delta e = e_a + c(t_r - t_c)$$

where e_a is the measured thermo emf, c represents change in voltage per degree difference in temperature $(t_r - t_c)$ and e is the corrected value of thermo emf.

A bridge type reference junction temperature compensator is shown at Fig. 5.15. The basic idea is to generate a voltage that will automatically compensate for any variations in reference junction temperature. Here is a thermally sensitive bridge supplied by a zener diode stabilised DC voltage source. R_c is a lead resistance compensator. The output voltage from the bridge (out of balance emf) varies with the enclosure temperature T_2, which is also the reference junction temperature (normally the ambient temperature), in such a way that the variations in cold junction temperature get nullified. Initially, by keeping both the junctions at ice point, the bridge is so adjusted that the output voltage is zero. This compensator takes care of moderate reference junction temperature fluctuations of ± 5°C and achieves a reference temperature stability of ± 0.2°C. Such a system introduces errors if the enclosure temperature undergoes wide variations.

A thermostatically controlled reference junction temperature (Fig. 5.16) is a still better compensator and gives better results.

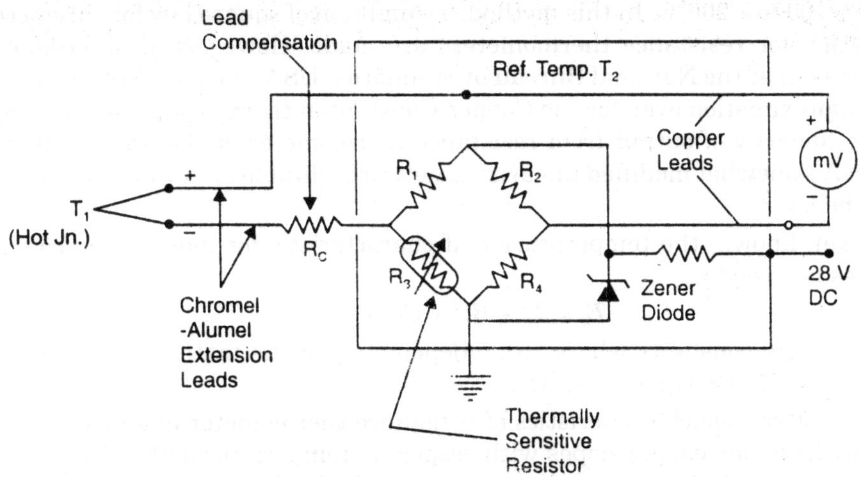

Fig. 5.15. Reference junction compensation.

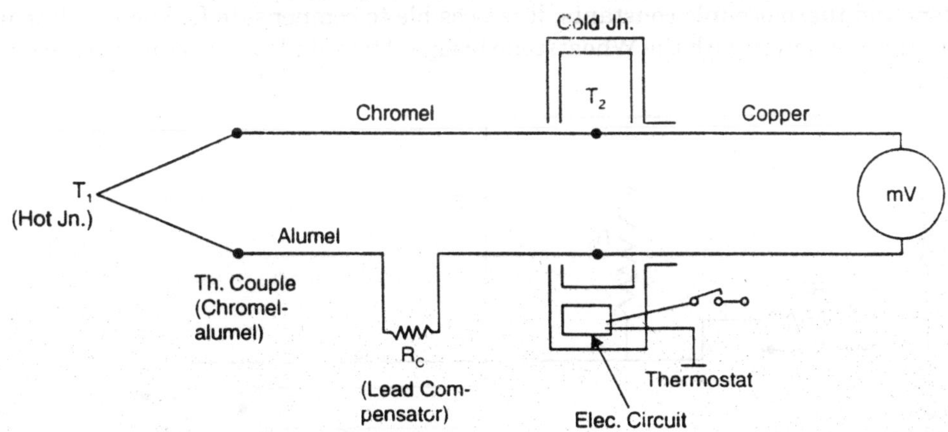

Fig. 5.16. Thermostatically controlled reference junction.

5.2.4. Linearization of Thermocouple Output

The output voltage-temperature characteristics of the commonly used thermocouples are generally non-linear which creates problems in designing control systems. There are some commercially available analog and digital devices "Diode function generator", "non linear digital to analog convertor" or the "Digital computer aided lineariser" etc. which are meant for linearising the output of various types of transducers. However, all these devices are complex and expensive. Simple networks using linear potentiometer have been devised and tried for obtaining a linear scale for the thermocouple output but these are again not quite suitable for digital control applications.

Yet another method uses a temperature sensing resistance in one arm of a Wheatstone bridge with one thermocouple as supply and another one in series with the bridge output for a temperature range from – 50 to + 50°C ; but this method does not take into account the load resistance.

In another scheme, two thermocouples in series with two temperature sensing resistances are used and connected in a differential mode in series, with a load resistance, for a range of temperature – 100 to + 200°C. In this method, assumption of square law for thermocouples and a linear relation for resistance thermometers are made. However, in accordance with the observations taken at the National Bureau of Standards, USA, the assumption of a square law is a broad approximation even for the Copper-Constantan thermocouple in the range – 100 to + 200°C. The linearity of output from resistance thermometer is also valid within the range 0-100°C only. A somewhat modified linearizing network suitable for a wider temperature range is suggested below :

As is well known, the temperature-emf characteristic for most of the commonly used thermocouples is given by :

$$E_t = At + Bt^2 + Ct^3 + \ldots\ldots \qquad \ldots(5.12)$$

where A, B, C... are constants whose value depends upon the material of the thermocouple, which can be determined experimentally.

The output voltage characteristics of resistance thermometer in a bridge with constant voltage supply have decreasing slopes with respect to temperature either Platinum or Nickel resistance element. If the constant voltage supply to the bridge be supplemented by a thermocouple, which is a temperature dependent voltage source, then with a proper selection of bridge parameters and thermocouple constants, it is possible to compensate for the non-linearities of resistance thermometers with the Wheatstone bridge. The simple linearising network is shown in Fig. 5.17.

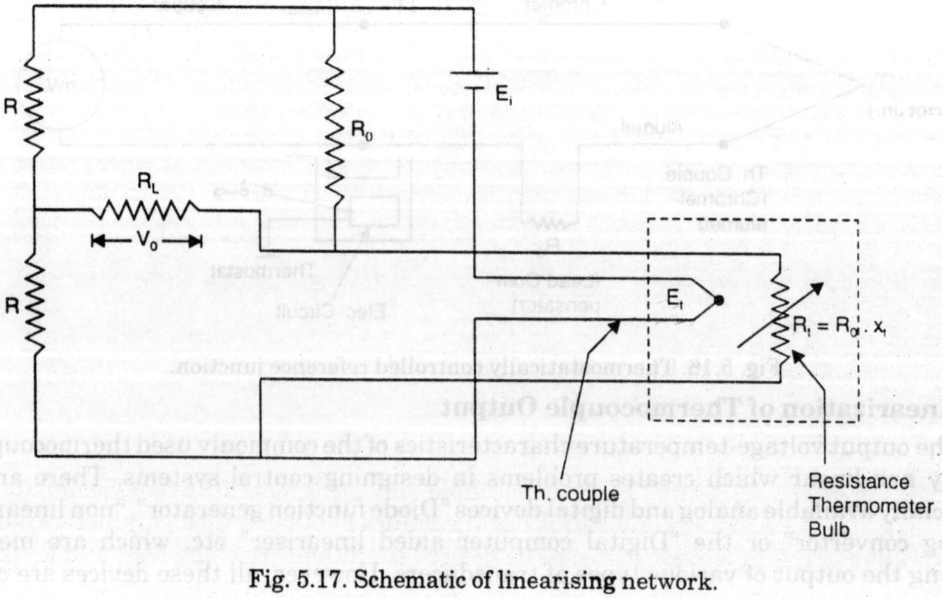

Fig. 5.17. Schematic of linearising network.

The slope of the temperature-resistance characteristics of the resistance thermometer can be given as :

$$x_t = R_t/R_o = 1 + \alpha t + \beta t^2 \quad ...(5.13)$$

where α and β are the temperature coefficients. A constant voltage source E_i in series with the thermocouple output voltage E_t represents the supply voltage of the modified bridge. The resistance of the constant voltage source and thermocouple lead wires are neglected. The thermocouple hot junction and the bulb of resistance thermometer form an integrated unit subjected to the same temperature, which is to be measured.

The output voltage across the load resistor R_L is given by :

$$V_o = \frac{(E_i + E_t)(x_t - 1)R_L}{ax_t + b} \quad ...(5.14)$$

where $a = 2R_o + R + 2R_L$ and $b = R + 2R_L$

The Eqn. 5.14, therefore can be rearranged as

$$E_i + E_t = \frac{V_o}{(x_t - 1)R_L}(ax_t + b) \quad ...(5.15)$$

Assuming a perfect linear output, $V_o = kt$

Where k is the assumed voltage sensitivity of the bridge in V/°C

Expanding the right hand side of Eqn. (5.15) upto the cube of t,

$$E_i + (At + Bt^2 + Ct^3) = \frac{2Ck}{R_L}\cdot\frac{1}{\alpha} + \left[\frac{at}{R_L} - \frac{2Ck}{R_L} - \frac{\beta}{\alpha^2}\right]t + \frac{2Ck}{R_L}\cdot\frac{\beta^2}{\alpha^2}\cdot t^2 - \frac{2Ck}{R_L}\cdot\frac{\beta^3}{\alpha^4}\cdot t^3 \quad ...(5.16)$$

where $C = (R_o + R + 2R_L)$

Comparing the coefficients of t on both sides of Eqn. (5.16), the condition for perfect linearity will by :

$$E_i = \frac{2Ck}{R_L}\cdot\frac{1}{\alpha} \quad ...(5.17)$$

$$A = \frac{ak}{R_L} - \frac{2Ck}{R_L} - \frac{\beta}{\alpha^2} \quad ...(5.18)$$

$$B = \frac{2Ck}{R_L} - \frac{\beta^2}{\alpha^3} \quad ...(5.19)$$

$$C = -\frac{2Ck}{R_L} - \frac{\beta^3}{\alpha^4} \quad ...(5.20)$$

Thus the relations between the different constants in case of perfect linear output are found to be :

$$A = \frac{ak}{R_L} - \frac{B\alpha}{\beta} \quad ...(5.21)$$

$$= \frac{ak}{R_L} - \frac{C\alpha^2}{\beta^2} \quad ...(5.22)$$

and

$$\frac{B}{C} = -\frac{\alpha}{\beta} \quad ...(5.23)$$

The constants A, B and C of thermocouple and the temperature coefficients α and β of the resistance wires are fixed quantities but the values of other bridge parameters, such as K, R, R_L and R_o can be adjusted to achieve the desired conditions. Keeping K, R and R_L fixed for a particular thermocouple, R_o is calculated from equation (5.14) by putting the value of E_i and solving for R_o.

Simplicity and advantages of Wheatstone bridge such as suppression of voltage or null output at any temperature are claimed to be fully preserved in the modified network.

BIBLIOGRAPHY

1. DVS Murthy, *Transducers for Instrumentation Systems*, Prentice Hall of India Ltd. 1995.
2. R.P. Singh, *Piezoelectric Polymers*, IETE Technical Review, 9(2), March-April 1992 p. 140-146.
3. W. Cady, Piezoelectricity, Mcgraw-Hill, New York 1946.
4. Marsh, R.H., *Selecting thermocouples and Platinum Resistance temperature detectors*, Control Engg. 18(11), Nov. 1971 p. 76-77.
5. Ashish Kumar Banerjee, *Linearization of Thermocouples*, Indian Journal of Pure and Applied Physics vol. 12, April 1974, pp. 303-306.
6. C.S. Rangan, GR Sarma and V.S.V. Mani, *Instrumentation : Devices and Systems*, Tata Mcgraw Hill 1983.
7. D Patranabis, *Sensors and Transducers*, Wheeler, 1997.

6

Ultrasonic And Magnetostrictive Transducers

Ultrasonic techniques have been employed for static and dynamic dimensional measurements either by finding the time taken for a pulse of ultrasound to move between two transducers attached to the opposite sides of a body (whose dimensions or thickness is required to be ascertained) or by measuring the difference in the echo reflection times from the opposite walls, using only one transducer.

Principles of ultrasonics, also enable us to find out the flow rate through a pipe or the average fluid velocity. This is achieved by measuring the difference in time that a pulse of ultrasound takes to travel upstream and downstream between two fixed points. The second wave propagated upstream will have a smaller effective velocity than the one propagated downstream. The difference in the two velocities will be exactly double the velocity of the moving fluid. This is called the "transit time" flow transducer.

Alternatively, the flow velocity can be computed by observing the change in frequency that occurs when an ultrasonic wave is scattered by a moving fluid. This frequency shift or Doppler shift of the scattered wave is directly proportional to the flow velocity.

The ultrasonic pulse echo technique is also used for monitoring liquid level. A sensor can be mounted on the outside wall of the tank monitoring the height of the back wall echoes which drops steeply when the liquid level reaches the sensor level. The method is useful for filling the tank to a desired level as well as for continuous monitoring of level.

The magnetostrictive transducers are, on the other hand, based on conversion of mechanical energy to magnetic energy and vice versa. Certain ferromagnetic materials, when mechanically stressed, undergo a change in its magnetic properties or conversely, when subjected to a change in the magnetic field, get their physical dimensions altered. The change in dimensions is proportional to the change in the applied magnetic field.

6.1. ULTRASONIC TRANSDUCERS

6.1.1. Displacement or Distance Measurements

If a short pulse of ultrasound is transmitted from one transducer, a pulse will be received at the transducer at the other end, after a time $t = d/c$ provided the velocity of ultrasonic wave in the medium is known and remains constant throughout the measurement (even if the distance d under measurement changes). Thus, the measurement of transit time enables the distance 'd' to be computed.

However, the waveform of the received signal is rather more complex since multiple reflections from the sides cause ultrasound signals to arrive after the directly transmitted pulse is received. Hence, it is advisable to measure transit time interval from the beginning of the transmitted pulse to the beginning of the first pulse that arrives at the receiving transducer.

This will produce a proportional output voltage. A bistable or a schmitt trigger will generate the square wave of period t, the transit time delay and an integrator (ramp generator) will give an equivalent sawtooth waveform of amplitude V (the output voltage) proportional to the period 't'. A peak detector and filter will detect the peak amplitude whose output will go to the recorder. This is shown in a block diagram at Fig. 6.1.

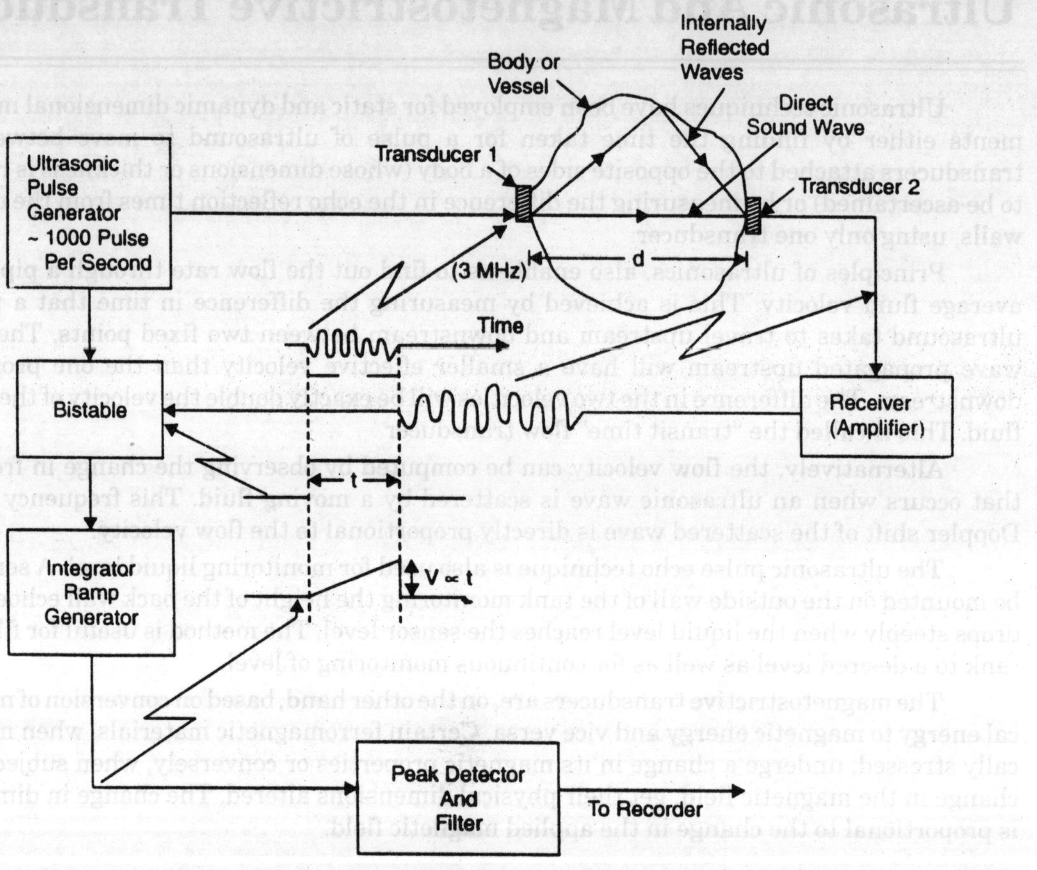

Fig. 6.1. Ultrasonic pulse echo technique.

The accuracy of this method is limited by the rise time of the sinusoidal burst of ultrasound and the dispersion of the pulse as it travels to the receiver amplifier. With a 3 MHz frequency, the rise time can be less than 0.5 μs ; for velocity of sound $c = 1.5 \times 10^5$ cm/sec., this corresponds to a distance of 0.075 cm. By increasing the frequency to 20 MHz, this resolution improves to 0.01 cm ; however, the greatly increased attenuation at this frequency also increases signal to noise ratio, especially at longer path lengths. The dynamic frequency response of the system is determined by the pulse repitition frequency. The method described above has been widely used for ultrasonic flaw detection in mechanical engineering and metallurgy.

6.1.2. Ultrasonic Flow Transducers

Ultrasonic flow measurement uses an ultrasonic wave transmitted through the fluid medium. The ultrasonic energy in the wave is coupled to the fluid through a piezoelectric crystal. A similar crystal is used as a receiver. For a crystal to transmit sound energy efficiently, its size has to be large compared to the wavelength of the transmitted wave. If the wavelength is of the order of 1.5 mm, the crystal dimensions could be about 15 mm. The velocity of sound through water is 1524 m/s ; the ultrasonic sensor therefore uses frequencies above 500 KHz. Again, there are two types adopted, one sensing the transit time through the fluid and the other is based on Doppler frequency shift.

6.1.2.1. Transducer Based on Transit Time

In Fig. 6.2 is shown a flow pipe carrying a liquid with an average velocity *V*. An ultrasonic source or transmitter and a detector or receiver are kept in the line of sight of each other. The emitted ultrasonic waves travel through the moving liquid and are received at the detector. With liquid stationary, the waves are transmitted at sound velocity *c* and therefore the travel time = L/c.

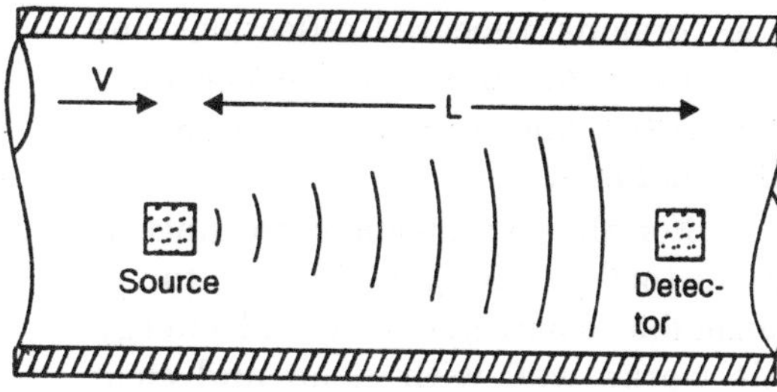

Fig. 6.2. Transit time based ultrasonic flow transducer.

The transmission speed increases with the liquid moving in the same direction.

Hence, travel time with fluid in motion = $L/(c + V)$ (*L* being the distance between the transmitter and detector)

The time difference $$\Delta t_1 = \frac{L}{c} - \frac{L}{c+V} \approx \frac{LV}{c^2}$$

Since $V << c$.

Keeping *L* constant for a given configuration, as long as *c* remains constant, Δt_1 is proportional to fluid velocity *V*. The presence of c^2 term causes a 2% change in Δt_1 for every 1% change in ultrasonic velocity *c*.

Interchanging the roles or transmitter and detector, the time difference

$$\Delta t_2 = \frac{L}{c-V} - \frac{L}{c} = \frac{LV}{c^2}$$

and $$\Delta t_2 - \Delta t_1 = \frac{L}{c}\left[\frac{1}{c-V} - \frac{1}{c+V}\right] = \frac{2LV}{c^2}$$

In practice, two transmitters and two receivers/detectors are placed as shown at Fig. 6.3, locations *A* and *B*. The waves transmitted by *T* at location *A* are received by the ultrasonic receiver *R* at location *B*. The two are linked through an external oscillator l. Likewise, the second transmitter-receiver set functions with the ultrasonic wave transmitted in the reverse direction through oscillator 2. The period of oscillation is decided by the transit time of wave. If t_1, t_2 be the transit times in the two cases, the two oscillator frequencies will be proportional to $(1/t_1)$ and $(1/t_2)$ respectively. Therefore

$$\frac{1}{t_1} - \frac{1}{t_2} = \frac{c + V\cos\theta}{L} - \frac{c - V\cos\theta}{L} = \frac{2V\cos\theta}{L}$$

This arrangement eliminates *c*.

The flow velocity sensed is the one integrated along the path of the ultrasonic wave. Whereas, it is normally required to know the velocity average across the cross section. The two average velocities are equal only for very large Reynold numbers *i.e.* turbulant flow when the velocity across the cross section is a constant. The error is maximum for laminar flow. The coefficient of discharge to account for this non uniformity is about 0.75 for Reynold numbers < 2000 (laminar flow). The coefficient varies from 0.93 to 0.96 for turbulant flow. In some versions, repeated reflections and reversal of signal paths are used between transmitter and receiver, in order to make up for this.

For the arrangement to function satisfactorily, the liquid has to be reasonable conductor of sound energy. This works well with clean and clear liquids without any suspensions or air bubbles. Turbidity or impurities in liquid affect not the functioning but accuracy.

The piezo crystals are usually embedded inside sections of piping with suitable flanges on either side. The unit as a whole can be installed permanently in the piping system.

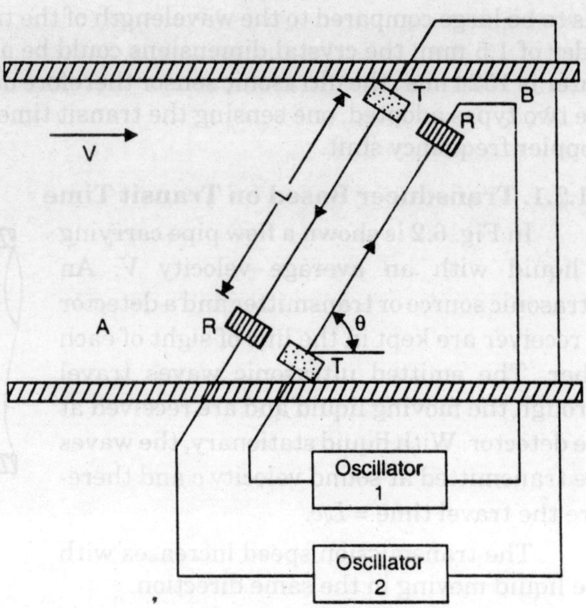

Fig. 6.3. Ultrasonic flow transducer arrangement avoiding dependence on ultrasonic velocity.

6.1.2.2. Ultrasonic Flow Transducers Based on Doppler Frequency Shift

The ultrasonic energy transmitted through a liquid gets partially reflected at every interface with particles suspended in the liquid, air bubbles etc. If the particle is in motion, the ultrasonic wave undergoes a shift depending on the velocity of the particles. The Fig. 6.4 (*a*) depicts an idealized situation. *T* and *R* represent the transmitter and receiver surfaces. An ultrasonic wave from surface *T* at point *A* is injected into the liquid at an angle θ (radians) with the flow axis. The peak of the wave originating at point *A* at a particular instant of time gets reflected at a particle surface at point *B* and returns to the receiver surface *R* at point *C*. The total distance travelled by the wave is therefore *ABC*. During the wave transmission, the particle moves with velocity *V* along the direction of flow. The next peak of the wave again emanates from *A* and continues upto the point *D* where it strikes against the same particle and is reflected. It comes to the receiver at point *F*. This peak traverses an extra distance of *BDE* compared to the last peak. It is therefore received at *R* with a time delay which is larger than the time period of the transmitted wave. The net result is an increase in the wavelength of the received peak and a simultaneous reduction in frequency. This frequency shift is called the Doppler shift and is a measure of the velocity of the particle suspension or the entrapped air bubble.

Let x denote the distance *ABC*

t_o the time of first peak originating at the transmitter t_1 the time of first peak arriving at the receiver.

Then $$t_1 = t_o + x/c$$

(*c* is the velocity of sound waves through the liquid)

Fig. 6.4. (*a*) Ultrasonic flow transduser based on doppler frequency shift principle.

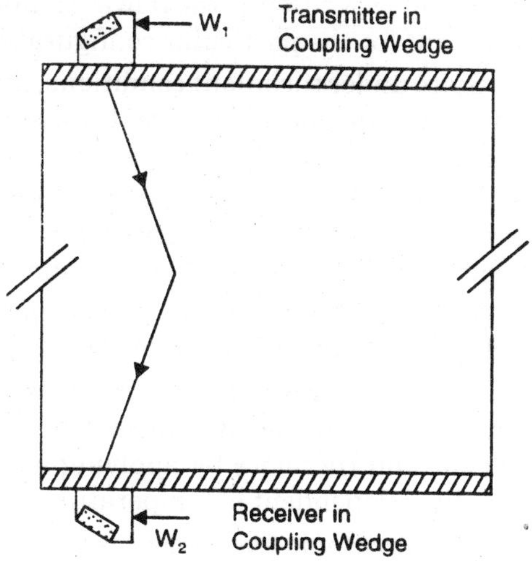

Fig. 6.4. (*b*) Ultrasonic flow transducer using wedges for coupling waves to the flowing fluid.

Likewise, if t_2 is the time of the next peak originating and λ the wavelength of ultrasonic wave,

$$t_2 = t_o + \lambda/c$$

The additional distance travelled by the second peak $BDE = 2v\lambda/c \cos\theta$

(v is the velocity of flow)

If t_3 be the time the second peak is received, then

$$t_3 = t_o + \frac{\lambda}{c} + \frac{2v\lambda}{c^2 \cos\theta} + \frac{x}{c}$$

The time difference between the two receiving instants

$$t_3 - t_1 = \frac{\lambda}{c}\left(1 + \frac{2v}{c\cos\theta}\right)$$

The corresponding received frequency $= \dfrac{c}{\lambda\,[1 + \{2v/(c\cos\theta)\}]}$

and the original transmission frequency $= c/\lambda$

The difference $$\Delta f = \frac{c}{\lambda}\left[1 - \frac{1}{[1 + 2v/(c\cos\theta)]}\right] \approx \frac{2fv}{c\cos\theta}$$

where $v << c$ and $f = c/\lambda$

Thus, the Doppler frequency shift Δf is directly proportional to the flow velocity of moving suspension particles.

This frequency shift can be converted into the equivalent analog signal by any FM demodulation technique,

In the equation $$\Delta f \approx \frac{2fv}{c\cos\theta}$$

the proportionality factor $2fv/c \cos\theta$ depends on :

ultrasonic frequency f,

the sound velocity through the flowing liquid which is temperature dependent and angle θ, which is a constant for the particular configuration.

Since under turbulant conditions, the velocity of flow across the cross section will be taken as same, the transducer measures the average velocity. By using a proper coefficient of discharge and the internal cross section area of the pipe, the volume flow rate can also be determined.

For satisfactory operation, a minimum suspension level is necessary, say 25 ppm of 30 micron size particles. This contrast with the transit time measurement system which requires suspended particles to be below a minimum, so as to ensure accuracy.

The particles are to be in suspension only above a threshold velocity (2 m/s for solid particles and 0.8 m/s for air bubbles).

The coupling wedges W_1, W_2 shown at Fig. 6.4 (*b*) also exert some pressure of their own in addition to the ultrasonic waves. The sound velocity for the wedge is much less temperature dependent and therefore by applying a small correction, the overall sensitivity can be made practically independent of temperature.

The ultrasonic flow sensor is quite robust and does not offer any obstruction to the flow. Installation and maintenance is easy.

The flow pipe diameter can be anywhere between 134 mm to 6 metre. Their operating temperature range is from – 40°C to 260°C. The liquid velocity can range.

From 0.3 to 30 m/s. An accuracy of 1% is normal and the scale is linear.

Since the piezo elements used have a limited size, the coupling of acoustic energy to gas as fluid is not practical. Hence the use of such transducers is essentially limited to sensing the flow of liquids only. Their cost is somewhat higher and ultrasonic sensors are yet to become popular.

6.1.3. Ultrasonic Level Sensor

As already indicated earlier in this chapter, this is a non contact type of sensor. The piezoelectric element mounted externally above the liquid surface emits ultrasonic waves. These get reflected from the liquid surface and also the container walls. The reflected waves are received by the piezo receiver and the time delay or the time of flight between the transmission and reception is computed from which the surface level is deduced. In order to ensure the effective coupling of the waves into air, the ultrasonic source has to be big in size. Although the ultrasonic wave gets focussed and more of it reflected from the walls, it causes widening of the received pulse and blurring of the signal. Since the received pulse shape changes with the liquid level, the templates of these received pulses for different levels are stored in a microcomputer during the calibration.

In actual operation, the received pulse is compared with the stored templates for matching and liquid level computed based on the best match. All this requires in situ testing and calibration. Considerable signal processing also adds to the cost. Because of these reasons, ultrasonic level sensing is less practical than flow velocity sensing.

6.1.4. Ultrasonic Transducers for Medical Applications

6.1.4.1. Diagnostic Imaging

Diagnostic imaging is the most widely practiced application of ultrasound. Real time grey scale ultrasound imaging has established itself as a non invasive, reliable and cost effective technique for providing images of internal body organs. The transducer sends short, focussed pulses of ultrasound into the body. It also receives the returning echoes and converts them into electrical signals which after amplification and processing are displayed for viewing on a high resolution monitor. The information content of the image is determined by the size and shape of ultrasound beam produced by the transducer and its efficiency in converting the returning echoes into electrical signals. The transducer therefore forms a vital part of an ultrasonic imaging system which greatly influences the image quality. An ultrasound image or frame is composed of a number of lines of echo information. The larger the number of lines, the greater the spatial resolution and the greater the amount of anatomic details capable of being displayed. However, more the lines required, the more time does it take to obtain a complete scan. This poses an important limitation for real time imaging, where the time available for an individual scan frame is short (less than 1/15 of a second).

Each line of echo information requires an adequate time for sound to travel from transducer to the deepest structures of interest and return. Thirteen microseconds are required for each centimetre of depth. All the above considerations can be related to the equation.

Frame rate X Lines per frame X Image depth = 76,923.

This equation states an upper limit on the combination of desirable features in imaging : temporal resolution, spatial resolution and field of view. Any increase in one parameter requires a decrease in the product of the other two. This restriction is imposed by the speed of sound in tissue.

In an ultrasound image, each line represents information from a slightly different direction. The returning echoes from each acoustic pulse are used to form one line. The direction in which the ultrasound beam is sent must be changed slightly to form the next line. There are two types of transducers in use depending on the method used to steer the ultrasound beam. These are called mechanical type and electronic type.

A. Mechanical Transducers

Here the direction of ultrasound beam is changed mechanically. As the image formed is generally sector shaped, these are popularly known as "Mechanical Sector Scanners or Transducers". These methods are used to steer the ultrasound beam produced by the piezoelectric crystal.

(*i*) Three crystals on a rotating drum. Each crystal becomes active as it comes opposite the acoustic window. [Fig. 6.5 (*a*)]. This popular design suffers from two drawbacks : It is technically difficult to connect electrically the fast rotating crystals ; other drawback is the relatively long dead time (during which none of the crystals is facing the acoustic window).

(*ii*) The ultrasound beam is allowed to sweep by using a single oscillating crystal also called Wobbler, [Fig. 6.5 (*b*)]. Electrical connections are also easy.

But the highest frame rate achieved is less than the other two varieties of mechanical transducers. Also, the changing angular velocity which is much higher in the central part and gradually reduces towards the ends of the sweep on both sides, making it difficult to have a uniform line density in the image. A uniform line density can be achieved by varying the pulse repetition frequency (PRF) but this increases the effective dead time of the transducer near the ends of sweep.

(*iii*) In this type of transducer the crystal assembly is fixed and an oscillating light weight sonic mirror is used to deflect the ultrasound beam. [Fig. 6.5 (*c*)]. The annular array type of mechanical transducers use this method of beam steering.

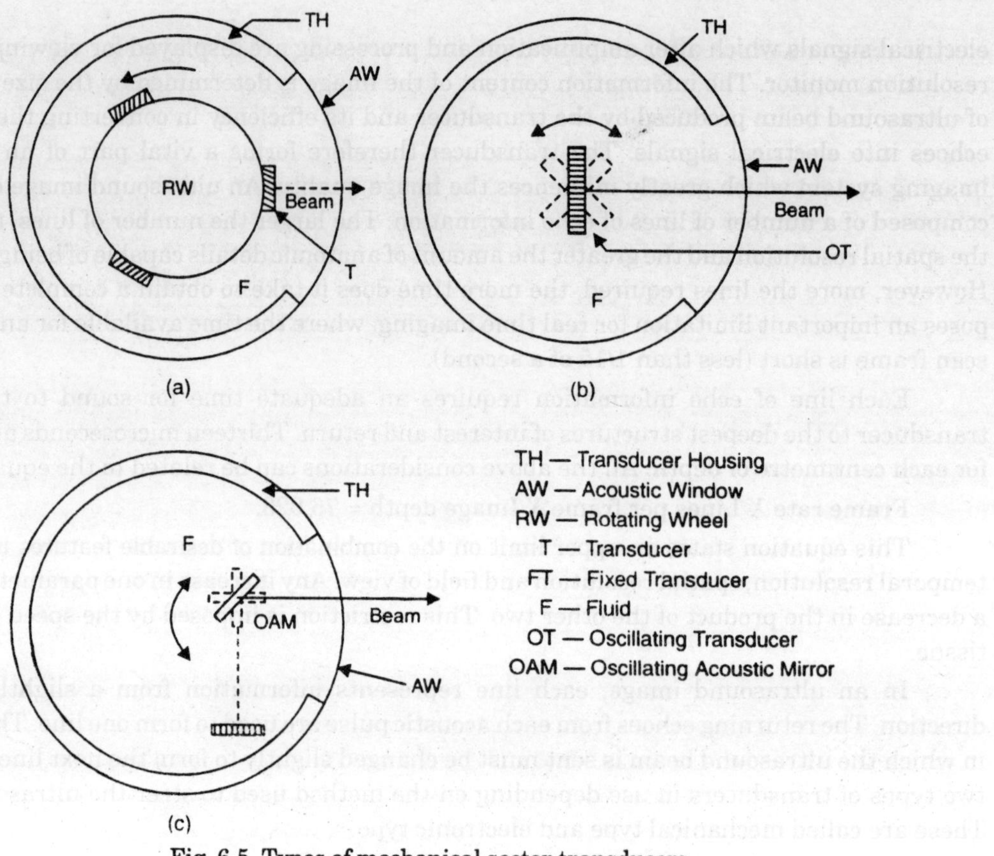

Fig. 6.5. Types of mechanical sector transducers.

Frame rates of 30 per second are easily achieved by mechanical sector scanners. Their principal advantages are simple design and less complex associated circuitry. However, the drawbacks are : limited life due to mechanical wear and tear, fixed focus leading to poor lateral resolution in regions away from the focal zone. This is overcome to some extent by annular array transducers using multiple concentric ring crystals. Ultrasonic beam can be focussed at various depths by such transducers. The presence of fluid in the transducer housing (necessitated by the presence of moving parts and acoustic impedance matching requirements). Air bubbles in this fluid are a frequent problem. Acoustic impedance mismatching between the crystal, the fluid and the transducer housing often lead to obscure the image in the near field.

B. Electronic Transducers

Electronic transducers employ a large number of small crystal elements. The sound beam is steered by either exciting different groups of adjacent elements or by using a sequential time delay in firing of all elements. Accordingly, there are two types ; linear sequential array and phased array. In both these types, the movements as well as focussing of the ultrasound beam is controlled electronically.

In a linear sequential array (LSA) transducer, a rectangular, parallel line field of view is produced. The array is comprised of many small piezoelectric elements arranged in parallel. The total length containing 100 or more elements may be 5 to 15 cm, each element being 1 mm or less wide. The beam produced by such a narrow element will diverge very rapidly after the wave front travels by a few mms. Thus, pulsing of a single element would result in poor lateral resolution due to beam divergence and low sensitivity due to small element size. To overcome this, several adjacent elements (8 to 16) are driven at a time. This increases the effective aperture and sensitivity. While moving to the next scan line, only one outermost element is dropped and added to the other side of the group. This gives an effective line spacing which approximates one element width. A significant variation to this is by mounting the piezo elements on a convex face rather than a linear surface. This provides a sector shaped image instead of a rectangular one. This design ensures a wider field of view at depth and a smaller contact area.

The phased array consists of 64 to 128 crystal elements closely placed together on the transducer face of 2-4 cm in size. All the elements are operated simultaneously but with a controlled phase relationship between them. This enables electronically sweeping the beam through all sector angles. The individual exciting pulses to the transducer elements get progressively shifted in time from each other by variable time delay modules before reaching the array, as shown in Fig. 6.6 (*a*). This shift in time is equivalent to a phase shift. The time shifts follow a linear pattern across the array from one side to the other. The uppermost element gets excited first and an approximately spherical acoustic wavefront begins propagating from it. The next element will be excited a very short time later and its wavefront will join the one from previous element. The sequence will continue till the last element is excited and the wavefront added to the total.

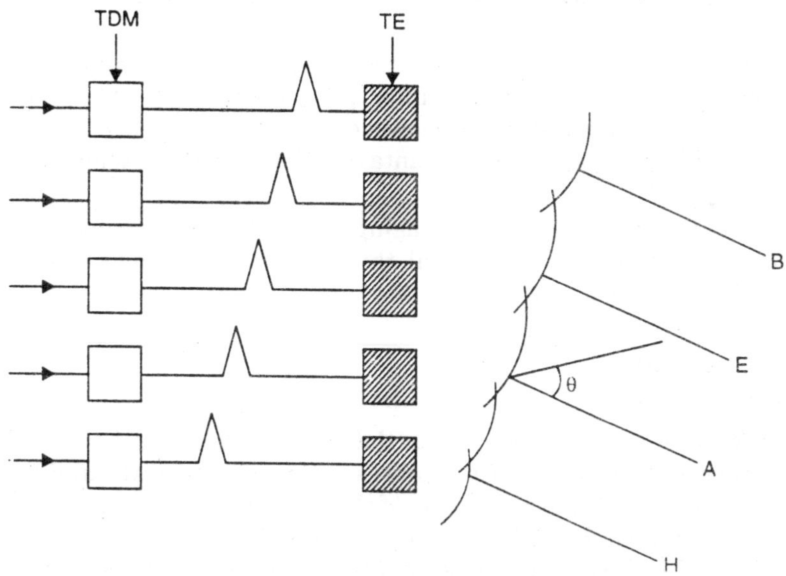

Fig. 6.6. (*a*) Electronic steering of ultrasound beam by phase shifting of pulses.

The combined wavefront which is sum total of the individual wavefronts gets tilted off at an angle θ away from the perpendicular. At this angle the individual wavefronts interfere constructively. At other angles there will be partial or total destructive interference. Thus, a substantial portion of ultrasonic beam propagates only in the direction of tilt determined by the phase shift between the piezo elements.

At certain angles, partial constructive interference occurs leading to the production of grating lobes which are peculiar to the multi element arrays and give off-axis sensitivity to a transducer used both as a source and receiver. Grating lobes artifacts can be reduced by using very narrow spacing between the elements and by employing very short acoustic pulses.

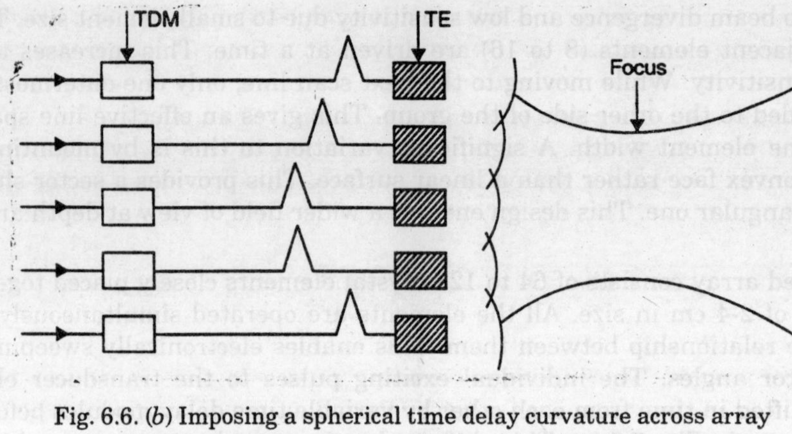

Fig. 6.6. (*b*) Imposing a spherical time delay curvature across array beam made to converge and focus at a desired depth.

Focussing of the beam in the plane along the direction of the transducer is LSA (linear sequenced array) and ESBPA (electronically steered beam phased array) is done by imposing a spherical time delay curvature across the array as shown in Fig. 6.6 (*b*). In ESBPA, simultaneous beam sweeping and focussing are possible. A unique advantage of the electronically focussed transducers is the ability to change the depth of focal zone simply by changing the magnitudes of the delays applied to the individual elements.

The absence of moving parts in electronic transducers make them more reliable and long lasting. Reflections from the transducer skin interface are further reduced by using multilayer quarter wavelength impedance matching. However, the relatively high cost of the ESBPA transducer slightly outweighs the above advantages. The electronic complexity is also quite high.

The quality of ultrasound diagnostic imaging ultimately depends upon signal processing and use of filters to soften the image. In general, three criteria are used for objectively assessing the information content of the image :

- Detail or spatial resolution
- Contrast resolution
- Image uniformity.

Detail or spatial resolution is regarded as the ability to distinguish two objects lying close together and is either axial or lateral. By axial resolution, we mean the ability to distinguish between two reflectors situated along the axis of the ultrasonic beam. It is determined by the pulse length, which is in turn dependent upon resonance frequency of the transducer and the damping employed. If 'c' is the average velocity of sound in the body tissues and 't' the effective pulse time, then axial resolution $AR = tc/2$

Again, if the extent of damping used is denoted by Q, called Q factor and 'f' is the resonance frequency, then

$$t = Q/2f$$

Hence $AR = Qc/4f$ or in terms of wavelength,

$$AR = Q\lambda/4.$$

Therefore, if follows that for better axial resolution or smaller value of AR, a transducer with lower Q factor (larger damping and smaller wavelength or higher frequency) should be used. It must be remembered that the attenuation of sound beam by the body tissues is frequency dependent. Higher the frequency of ultrasound, more is the attenuation, leading to reduced tissue penetration. It is advisable to use the highest ultrasound frequency consistant with adequate tissue penetration, for best *AR*.

Since the ultrasound beam consists of a band width of frequencies around the resonance frequency, therefore, as higher frequency ultrasound is absorbed more by the body tissues, the sound beam reaching deeper areas has a lower mean frequency. This leads to slightly poorer axial resolution with increasing depth.

Another way to improve *AR* is by lowering the *Q* factor. This is accomplished by using a damping material, such as Aluminium or Tungsten filled epoxy on the back of the crystal. Similar effect can be produced by supplying reverse bias voltage to the vibrating crystal.

The lateral resolution is the capability of resolving two reflectors in a direction prependicular to the ultrasonic beam and is strongly dependent on depth of penetration. Two objects placed side by side can only be detected as separate when the beam width at that depth is less than the distance between them.

Therefore, the narrower the beam, the better is the lateral resolution (*LR*) at a given depth.

A typical unfocussed ultrasound beam produced by a circular disc transducer is shown at Fig. 6.7 (*a*). It consists of a parallel component, which is the Fresnel or near zone, and a diverging component called Fraunhoffer or far zone. There is a transition point in between the two length of the near zone or the transition distance '*T*' is determined by the transducer disc diameter '*D*' and the wavelength 'λ' and is given by the relation

$$T = D^2/4\lambda$$

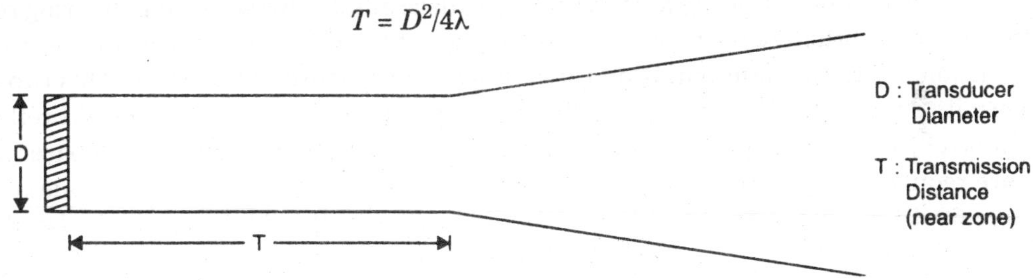

Fig. 6.7. (*a*) The beam diverging away from an unfocussed disc transducer.

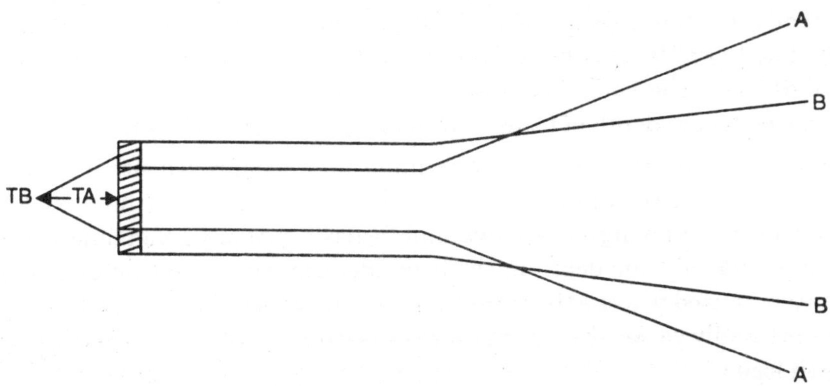

Fig. 6.7. (*b*) The beam from a small transducer (*TA*) diverges more rapidly than larger one (*TB*).

Focussed beams are used for obtaining the maximum resolution. Focussing is only possible in the near zone and the beam will be divergent in the far zone regardless of the focussing used. The transducer diameter and the frequency should be so chosen that the region of interest lies within the near zone. The angle of divergence of beam in the far zone is inversely proportional to the transducer diameter. The ultrasound beam from a smaller transducer may be narrower at the start but gets wider as compared to a beam from a larger transducer [Fig. 6.7 (*b*)]. This is because of a shorter near zone and a greater angle of divergence of a smaller transducer.

In single crystal transducers, the ultrasound beam is focussed either by having it of curved shape or by placing an acoustic lens in front of it. The diameter of the focal spot '*d*' can be calculated from the realtion :

$$D = 2.44\, F\lambda/D$$

where *F* is the focal length of the lens and '*D*' the transducer diameter. Thus for better *LR*, shorter focal length and shorter wavelength (or higher frequency) and larger transducer diameter are required.

However, there is one disadvantage of having short focal length. The *LR* at the depths away from focal plane gets hampered. The axial distance over which the beam maintains its approximate focussed size, is termed the depth of focus and is given by :

$$\text{Depth of focus} = 3(F/D)^2\, \lambda$$

This equation indicates that the requirements for better *LR* and larger depth of focus are mutually contradictory and a compromise in the design of acoustic lens is called for.

The Contrast Resolution (*CR*) is the ability to discriminate between neighbouring regions of differing image brightness. It permits detection of subtle differences in low level echoes and is responsible for tissue differentiating capabilities of an ultrasound scanner. This capability also permits detection of low level echoes from soft tissues in the presence of bright reflectors. *CR* can be objectively assessed by examining the beam plots of a transducer. Separate beam plots are required for different depths. The weaker echoes that determine the tissue textures are 40 to 60 dB below the brightest (zero dB) echoes. The width of beam plot at – 50 dB level gives an indication of *CR*.

Image Uniformity

This parameter of image quality deals with the uniformity of *LR* and *CR* over the field of view. As already mentioned both *LR* and *CR* are depth dependent. The mechanical sector scanners using fixed focal length transducers provide good resolution in the focal plane. Still there is loss of detail in the near filed and blurring of image in the far field. Annular array transducers provide longer depth of focus improving the image uniformity.

In phased array transducers, the image uniformity is improved either by using multiple transmit foci, thereby effectively producing a narrow beam from near to the far field, or by using a continuously variable receiving focus or dynamic focussing. At any given time, echoes are being received from one and only one depth in the body (determined by the velocity of sound). This timing information is used to focus the transducer to receive echoes from the proper depth. This receiving focus is rapidly varied (by a computer software control) so that echoes returning from all depths are in focus.

6.1.4.2. Focussed Ultrasonic Transducer for Disintegration of Kidney and Gall Bladder Stones

The kidney stones are mainly composed of Calcium Oxalate, Magnesium Ammonium Phosphate and Urates, whereas gall bladder stones are cholesterol and bilirubinate. These are required to be disintegrated through surgical operations. The newer ways of doing so with focussed acoustic energy, based on ultrasonic sources comprising piezoelectric transducers have been designed to meet the demands of this particular application area.

A therapeutic focussed transducer using PZT-4 piezoelectric disc has been designed and fabricated at National Physical Laboratory, New Delhi. This is shown by a sketch in Fig. 6.8. The PZT-4 disc is mounted on the inner wall of a spherical Aluminium housing and fixed by a cement Adhesive or epoxy resin. The front surface thickness of transducer housing has been controlled by machining to λ/4 *i.e.* one fourth of the wavelength of Aluminium, because it allows less absorption of ultrasound. A specially designed metal spring is used on the back surface of the transducer disc which also works as one of the electrodes. The second electrode is kept at the front surface in contact with the metal casing. There is air backing (low impedance layer) to get optimum transfer or acoustic energy. The lateral resolution of all transducers is obtained by keeping minimum bandwidth which is achievable by reducing the transducer element. In this case, 1 MHz disc is used to achieve good intensity at the maximum depth of resolution. Such single element transducers are used together in the form of an array. Seven transducers, separated by equal distance are centred onto a metallic semi-sphere. Each transducer has a maximum power handling capacity of about 14.8 W. Seven transducers in an array transmit through focussing, enough power to break the stones. Each transducer is individually excited through a high power (upto 150 W) ultrasonic generator. The power is limited to 18 W across each transducer due to the limitation of PZT-4 material used here.

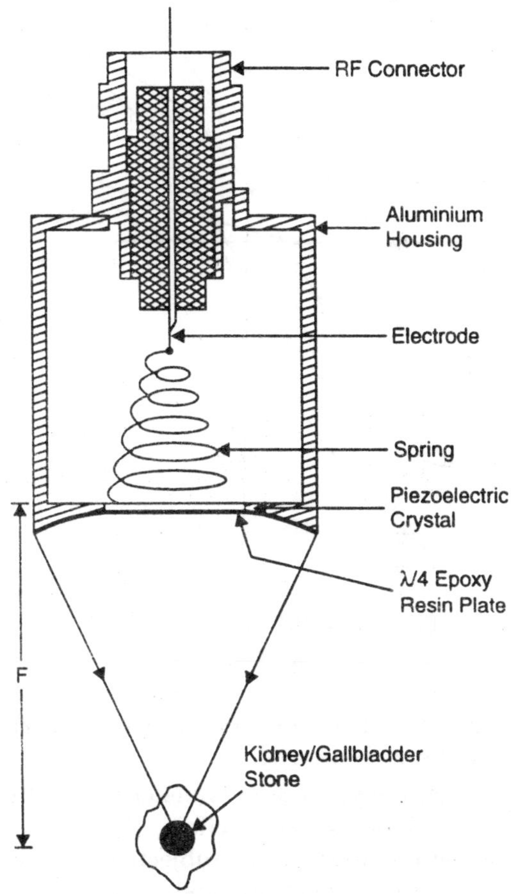

Fig. 6.8. Schematic of a focussed ultrasonic probe for lithotripter.

When focussed waves come in contact with an interface, the sound impedance changes, so that the compression phase or tensile phase is reflected which further depends on the acoustic quality of this interface. At the point where the focal wave exceeds the strength of the material, mechanical destruction occurs. An externally generated focussed ultrasound wave enters the body tissue and the water. The renal calculi and gall bladder stones are exposed to ultrasound beam with 50-60 W/cm^2 intensity for a particular time interval depending upon the size of the stone. Higher intensity is required to break kidney stones.

6.2. MAGNETOSTRICTIVE TRANSDUCERS

As already mentioned earlier in this chapter, magnetostrictive transducers are converters of mechanical energy to magnetic energy and vice versa. These transducers are rugged in construction and have low output impedance. Magnetostriction is shown by Nickle and Nickle-Iron alloys. Since the magnetic properties are normally represented by *B-H* curve, it is the non linearity of the *B-H* characteristic which limits its scope of application, especially when high accuracy is desired.

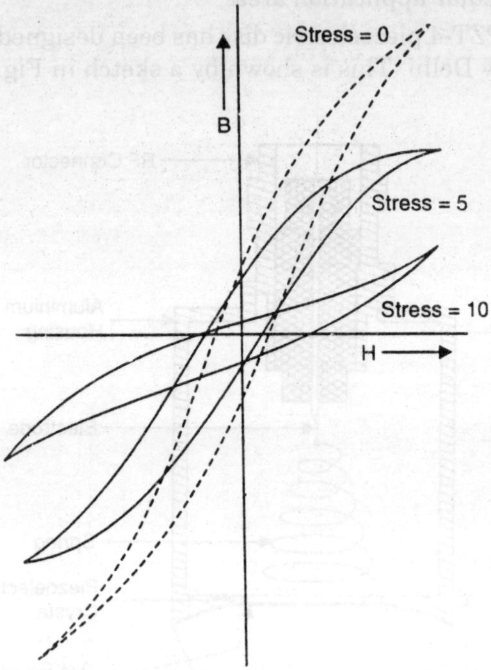

Fig. 6.9. (*a*) *B-H* loops for Ni with varying stress levels.

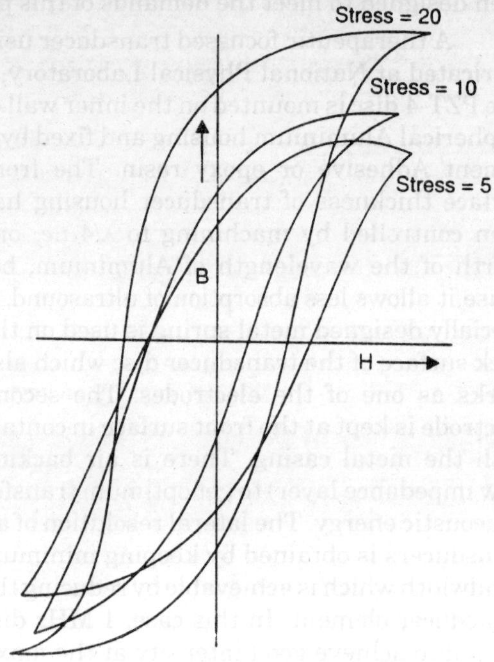

Fig. 6.9. (*b*) *B-H* loops for Ni-Fe with varying stress levels.

Generally, noise in transformer is due to the reverse magnetostrictive effect, that is, when ferromagnetic materials are subjected to a change in magnetic field, their physical dimensions tend to alter, thus producing a noise.

The major drawback of these transducers is that each transducer needs individual calibration before use. This is because the magnetic properties of materials differ from one another, thus leaving no scope for inter changeability.

- As indicated by the Fig. 6.9, the shape and size of the *B-H* loop gets sufficiently altered when the ferromagnetic material is subjected to tensile, compressive or shear stress.

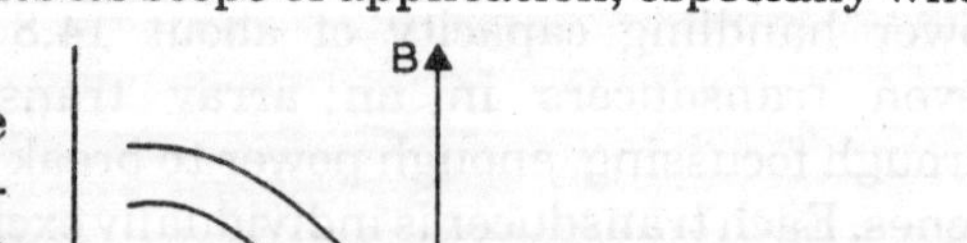

Fig. 6.9. (*c*) Stress *Vs B* curves for Ni with increasing values of *H*.

Permeability increases with the tensile stress increasing (positive magnetostriction) in case of Ni-Fe alloys while it decreases with the increasing tensile stress (negative magnetostriction) in case of Ni. In fact, it is the percentage of Ni that considerably affects the variation in permeability.

Ferromagnetic materials exhibiting magnetostriction are sensitive to the polarity of stress, hence such transducers enable us to measure alternating forces.

However, magnetostrictive transducers offer no competition with the piezoelectric transducers, because the effects of stress on permeability and remanance are non linear and exhibit hysteresis. Such transducers also require a mechanical bias so as to make them suitable for the application of cyclic stress.

6.2.1. Magnetostrictive Force Transducer

The self inductance of an iron cored coil changes if the core characteristic is changed due to the application of force. The mechanical strain affects the orientation of magnetic domains and this results in a change in effective permeability. To obviate this, the core is to be laminated and the magnetic path should be continuous with no air gaps present. Current through the coil is so adjusted as to make most sensitive to stress. 10-20% change in inductance L is observed with Ni-Fe alloy transducers. Permeability of the core material increases when the stress is compressive and decreases when the stress is tensile. A sketch is shown in Fig. 6.10. While it is difficult to attain high accuracy of measurement with this type of transducer, ruggedness is an advantage. Measurement of large static forces is possible with an accuracy of 2.5% F.S.

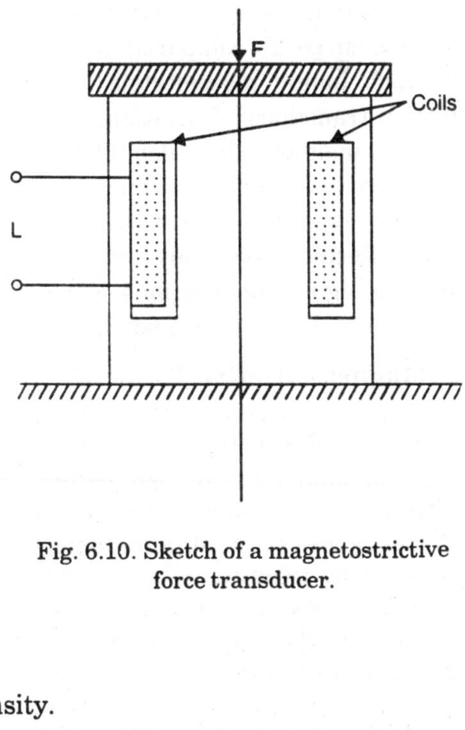

Fig. 6.10. Sketch of a magnetostrictive force transducer.

For dynamic force measurements, core is to be polarised with DC mmf so that the operating point of flux density is held at some convenient point below saturation. Application of stress 'S' then results in a change B to $B \pm \Delta B$, depending upon the material. The sensitivity of transducer is then given by :

$$\Psi = \frac{\Delta B}{S}\Bigg|_{B = B_o}$$

where $B = B_o$ denotes the operating point of flux density.

For small sinusoidally varying values of S, the variation ΔB can also be taken as changing sinusoidally. Sensitivity Ψ is maximum in case of Ni-Fe alloy (with Ni = 68%, Fe = 32%), when B_o is adjusted to $1/\sqrt{3}$ of saturation flux density. For reasons of simplicity and stability, B is chosen as equal to remnant flux density. Fall in sensitivity can partly be made up by providing more number of turns on the pick up coil. Then, the emf induced in the winding

$$e(t) = \Psi AN \frac{ds(t)}{dt}$$

where 'A' is the cross sectional area of the coil and 'N' the number of turns on it.

The transient forces and stresses can be measured by integrating $e(t)$ with respect to t.

6.2.2. Magnetostrictive Acceleration Transducer (Addition of a Proof Mass)

The force transducer can be converted into an acceleration transducer by adding a proof mass (provided by a brass cylinder) which has a mass equal to the core itself. A flexible diaphragm supported above the brass cylinder, guides the transducer against responding to the transverse accelerations. Induced emf in the coil is integrated over time. Bandwidth is extended to lower frequencies. However, this type of transducer has not become popular because the entire size and weight is rather too large as compared with very much handy piezoelectric acceleration transducer. Variations in the earth's magnetic field also affect its sensitivity. Core laminations are to be rigidly held in position so as not to be affected adversely while measuring higher accelerations. Fig. 6.11 gives the details of transducers.

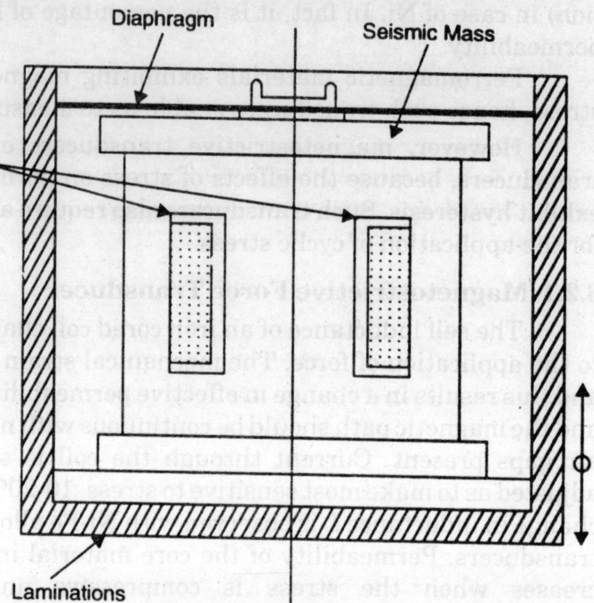

Fig. 6.11. Magnetostrictive acceleration transducer.

6.2.3. Magnetostrictive Torsion Transducer

This is a specific type of transducer and as illustrated in Fig. 6.12, it consists of a Ni wire 0.5-1.0 mm diameter, stretched between the poles of a permanent magnet and has a small stylus rigidly attached to it at mid-point. Ni wire is prestressed by twisting it before being installed into position. Two pick-up coils of fine wire are wound around the wire on either side of the mid-point. Any displacement of stylus to one side or the other increases the torsion on one side and decreases it by an equal amount on the other side. This results in an increase of magnetic flux in one half and decrease in the other half. The corresponding induced emfs are in phase opposition and processed suitably in LVDT. It is used as a phonograph pick-up and designed to have a flat frequency response over the range 150 Hz-15 KHz. Due to non linearity and hysteresis in performance, it normally finds limited application in such situations where time varying torsions of small amplitude are to be measured.

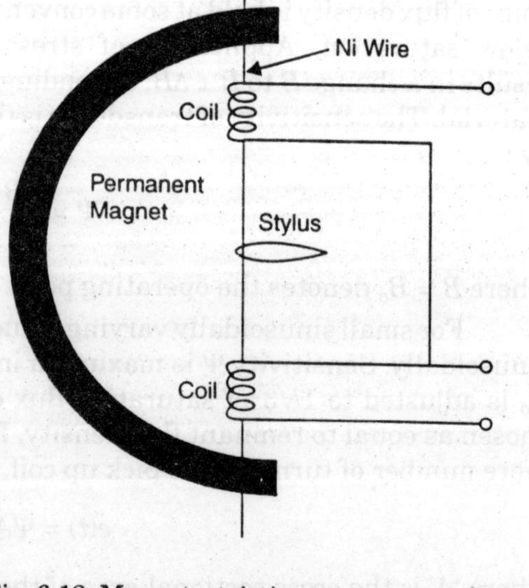

Fig. 6.12. Magnetostrictive torsion transducer.

BIBLIOGRAPHY

1. DVS Murthy, *Transducers and Instrumentation Systems*, Prentice Hall of India Ltd. 1995.
2. Richard SC Cobbold, *Transducers for Biomedical Measurements : Principles and Practices*, John Waley & Sons New York 1974.
3. TR Padmanabhan, *Industrial Instrumentation : Principles and Design*, Springer-Verlag London 2000.
4. Ashok Kumar, *Ultrasonic Sensors*, Paper presented and published in the Proc. of National Conf. On Sensors and Transducers, Chandigarh, Sept. 20-21 1996, p. 86.
5. BS Verma, *Ultrasound Transducers and Image Quality*, IETE Technical Review, 9(2), March-April 1992, p. 128.
6. Ravinder Agarwal and VR Singh, *Focussed Ultrasonic Transducer System for Disintegration of Renal / Gall Bladder Stones*, IETE Technical Review 9(2), March-April 1992 p. 133.
7. Anil Suryavanshi, VR Singh and Nawab Singh, *Ultrasonic Arrays for Medical Diagnosis and Structural Analysis of Materials*, IETE Technical Review, 9(2) March-April 1992, p. 136.

7

Electrochemical Transducers And Ion Electrodes

Electrochemical sensors and transducers that are sensitive to ions of a specific type and those measuring the effective ion concentration have been widely used for medical as well as for industrial purposes. For example, glass electrodes are adapted to form a standard method for measuring the hydrogen ion concentration. Their adaptability for in vivo measurements at the cellular level is a unique property attributed to these electrodes.

The measurement of hydrogen ion activity as also the partial pressures of oxygen and carbon dioxide in blood enable us to discover the nature and origin of disturbances to the acid-base equilibrium in human body.

7.1. BASICS OF ELECTRO CHEMISTRY

A study of fundamentals of electrochemistry is essential to understand the basic principles of electrochemical transducers and the way how the ion selective electrodes operate.

7.1.1. Properties of Electrolytes

Free electrons are not present in the electrolyte solution ; but these move as the integral parts of ions and molecules. As is well known, pure water is a very poor conductor of electricity and gets ionized into hydrogen ion (in hydrated form) and hydroxyl ions. The presence of a free ion in aqueous solution causes the surrounding water molecule, to get attracted to the ion as shown in the Fig. 7.1 (*a*) and (*b*). Most of the properties of ions are controlled by the degree of hydration. Ionic mobility is one such property which gets greatly reduced by the viscous drag resulting from the hydration shell. In weak electrolytes like acetic acid, the variation in dissociation of the substance causes a change in conductivity with dilution.

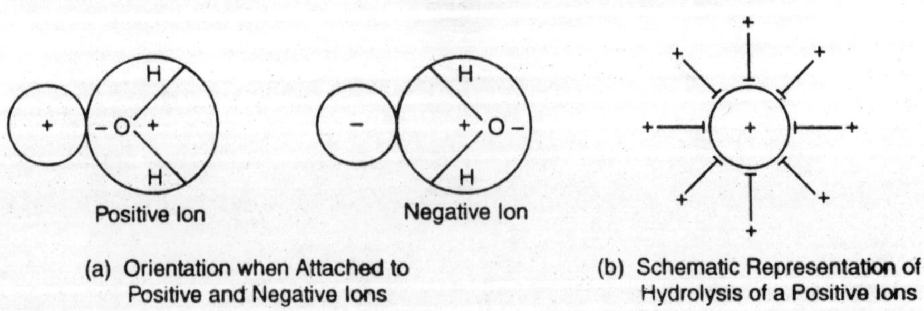

Fig. 7.1. Structure of the water molecule.

In a strong electrolyte like KCl solution, the K^+ and Cl^- ions will mutually interact and in the process get surrounded by each other. As positive K ions move to the negative electrode, the surrounding Cl^- ion cloud moves in the opposite direction, thereby creating a space charge that tends to restore the ions to their original positions. This results in reduction in the transport speed of ions. Hydration of Cl^- ions causes the K^+ ions being subjected to an increased viscous drag. The inter-ionic forces thus decrease the conductance of the electrolyte. For large dilutions when the interactions are negligible, the conductance will vary with the degree of dilution, in a predictable manner. The inter ionic effects are also of importance in determining the availability of an ion to participate in a chemical reaction. Ionic activity could be defined as the concentration of ions which, in the absence of interionic effects, would produce the same effects as the real solution.

Thus the ionic activity 'a' can be expressed as :

$$a = \gamma c$$

where c is the concentration and γ the activity coefficient. $\gamma = 1$ for an infinitely dilute solution and then the activity becomes equal to concentration. In a simple electrolytic solution with a single univalent anion and cation, the mean ionic activity a_m is related to the mean ionic activity coefficient γ_m by

$$a_m = \gamma_m c \qquad ...(7.1)$$

If the electrolyte anions and cations have activities a_{m1} and a_{m2} and the corresponding activity coefficients γ_{m1}, and γ_{m2}, respectively then

$$a_{m1} = \gamma_{m1} c$$

$$a_{m2} = \gamma_{m2} c$$

Multiplying and taking square root,

$$\sqrt{a_{m1} \cdot a_{m2}} = \sqrt{\gamma_{m1} \cdot \gamma_{m2}}\; c \qquad ...(7.2)$$

Comparing equations (7.1) and (7.2) :

$$a_m = \sqrt{a_{m1} \cdot a_{m2}} \quad \text{and} \quad \gamma_m = \sqrt{\gamma_{m1} \cdot \gamma_{m2}}$$

The variations of the activity coefficient with ionic concentration is given by the following equation :

$$\log(\gamma_m) = \frac{-0.51\, z_1 z_2 \sqrt{\mu}}{1 + 3.3 \times 10^7\, a_i \sqrt{\mu}} \qquad ...(7.3)$$

This equation is supported by theoretical and experimental work and holds good at a temperature of 25°C. Here z_1 and z_2 indicate the valences of cation and anion respectively, a_i is the effective diameter of the hydrated ion in solution and μ the total ionic strength, expressed as the sum over all ions in the solution, is given by the relation

$$\mu = \frac{1}{2} \Sigma\, z_i\, c_i \qquad ...(7.4)$$

where c_i is the concentration of the ith ion of the valence z_i. The mean activity coefficient of a given electrolyte is approximately constant in all the solutions having the same total ionic strength. This is true for a dilute solution.

7.1.2. Half Cells and their Potentials

When certain metal electrodes are immersed in a solution containing their ions, an equilibrium is quickly established whereby the rate at which the metal atoms lose electrons and pass into the solution is exactly balanced by the rate at which metal ions in solution deposit on the electrode as metal atoms.

During the process of establishing thermodynamic equilibrium, a charge redistribution occurs in the immediate vicinity of the metal-solution interface and establishes an equilibrium potential difference between the metal and electrolyte. For a potential to exist under equilibrium conditions, it is necessary that there be one or more regions, where the net charge density is non zero. Such a region called the space charge region, exists in the immediate vicinity of the metal-solution interface.

The space charge region consists of a compact layer of charge formed by the ions at the closest distance of approach to the interface and a more diffuse layer, which extended out much further from the metal-solution interface. The compact layer is subject to much greater electrostatic forces and is more securely bound than the diffuse layer. Thickness of the compact layer is roughly equal to the radius of an hydrated ion, while the diffuse layer, in a dilute solution, may be of the order of 50 Å. The thickness of the diffuse layer decreases with increasing concentration. This is shown in Fig. 7.2.

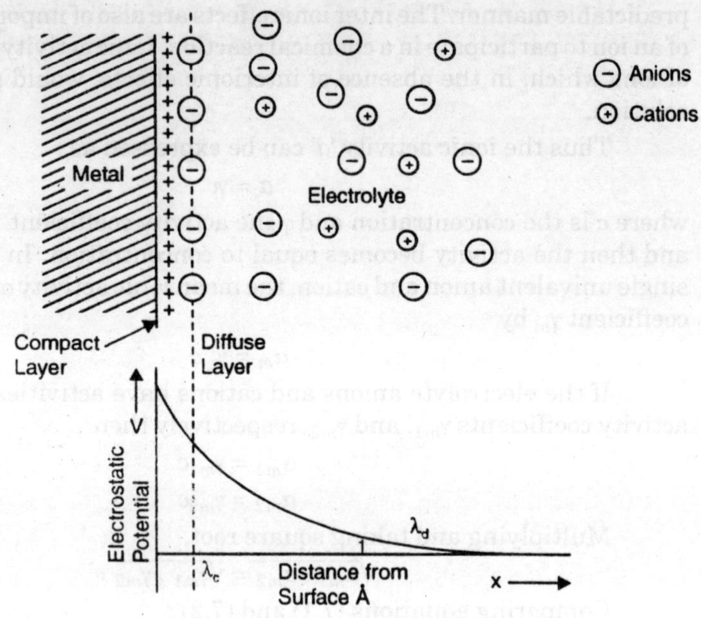

Fig. 7.2. Structure of the double layer (compact layer of charge and diffuse layer).

All metals do not reach an equilibrium with water. Some metal, like sodium react with water until the electrode is fully consumed. Certain noble metals such as gold and platinum are the cases where ion exchange reaction with water is normally dominated by impurities rather than by the intrinsic properties of the metal itself. In order to measure the potential, a second metal electrode is required in the solution to complete the electrical circuit. This second electrode itself generates a metal-electrolyte potential so that the potential difference between the two electrodes could be measured. This requires that one of the electrodes should generate zero potential so as to measure the potentials of all other electrodes with reference to it. By an international agreement, the hydrogen electrode when operating in a specified manner, has been assigned a potential exactly equal to zero.

The standard hydrogen electrode consists of a platinum electrode with an electro-deposit of platinum black, immersed in an acid. If the hydrogen ion activity of the acid is 1 mole/litre (pH = 0), and hydrogen gas is bubbled through the solution around the electrode so that its partial pressure is 1 atmosphere, the potential is defined to be zero at all temperatures.

Ignoring the hydration of the hydrogen ion, the reaction at electrode can be represented as :

$$H_2 \text{ (aqueous solution)} \rightleftharpoons 2H \text{ (absorbed on metal)}$$
$$\rightleftharpoons 2H^+ \text{ (aq. solution)} + 2e^-$$

Thus, the platinum electrode acts as a catalyser of equilibrium as well as a supplier of electrons (negative charge), without itself reacting in the manner of a metal-ion electrode.

The potential generated by the hydrogen electrode is called redox (oxidation-reduction) potential. This could be defined as a potential generated through reactions in which two phases exchange electrons and for which the electrode serves only as a means for this exchange.

7.1.3. Electrochemical Cells

An electrochemical cell consists of two half cells with their electrolytes in electrical contact. The emf of the cell depends on the difference in the two half cell potentials as also on the liquid junction potentials that may be present.

For a single Daniel Cell shown in Fig. 7.3 (*a*), if the two solutions have metal ion activities of unity, it can be represented as :

$$Zn/Zn^{2+}\ (a = 1)\ |\ Cu^{2+}\ (a = 1)/Cu$$

Here, the single oblique line indicates the phase boundary and the vertical broken line represents a membrane permeable to all ions to flow from one solution to the other but at the same time prevent the free mixing of the two solutions.

The equilibrium potential of the cell is given by the Nernst equation

$$E_c = E_o + \frac{RT}{2F} \ln\ (a\ Cu^{2+}/aZn^{2+}) \qquad ...(7.5)$$

where R is the gas constant, F the Faraday's constant, T the absolute temperature and E_o the standard electrode potential in volts (available from the standard reference tables). The tables give standard potential (at 25°C) for

$$Cu^{2+}/Cu = +\ 0.337\ \text{V and that for } Zn/Zn^{2+} = -\ 0.763\ \text{V}$$

Because $$E_o = +\ 0.337 - (-\ 0.763) = 1.100 \qquad ...(7.6)$$

Hence $$E_o = 1.100 + \frac{RT}{2F} \ln\ (aCu^{2+}/aZn^{2+}) \qquad ...(7.7)$$

The above expressions assume the potential between the two solutions to be negligibly small. In practice, this may not be the case. To eliminate such potentials, a salt bridge is used as illustrated in Fig. 7.3 (*b*). This normally consists of a tube filled with a saturated or highly concentrated solution of KCl. The two ends of the bridge solution are in electrical contact with the two solutions to be joined but are plugged sufficiently so as to prevent excessive mixing. The

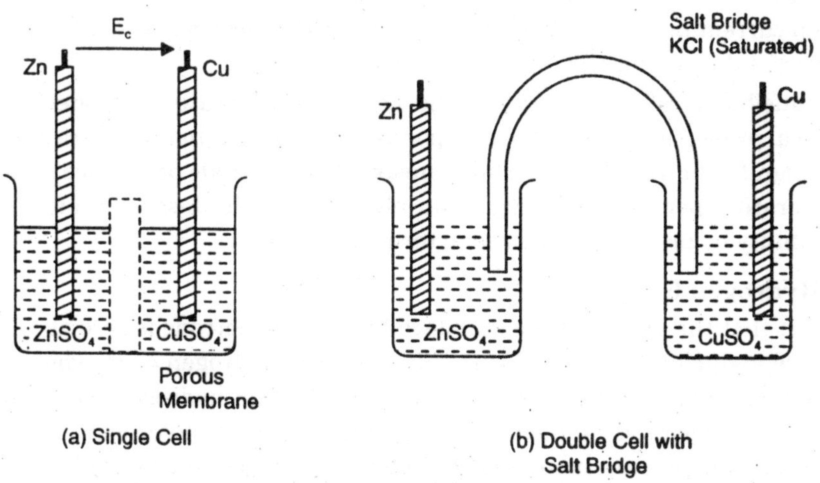

Fig. 7.3. Daniel electro-chemical cell.

two liquid-liquid junctions generate only a very small emf of a few millivolts only, often negligibly small in comparison to the cell emf.

When a salt bridge is used and the liquid-liquid junction potentials are neglected, the cell is represented by

$$Zn/Zn^{2+} \;//\; Cu^{2+}/Cu$$

where double slash indicates the elimination of the liquid junction potential.

7.2. THE REFERENCE ELECTRODES

The electrochemical measurement of ion activity basically involves two electrodes, a measuring electrode and a reference electrode. It is the measuring electrode which acts as a sensor for the quantity being measured and should therefore have quick response to the variations in the ion activity as well as reproducibility.

The function of the reference electrode is to provide electrical continuity with the test solution in such a manner that its potential is independent of any changes in the test solution. Since all electrodes are temperature dependent, the measuring and reference electrodes should be so designed that their temperature coefficients cancel out each other.

When the measuring electrode is an ion responsive glass electrode, two reference electrodes are normally used. An inner reference electrode immersed in a suitable standard electrolyte provides electrical contact to the inner surface of glass electrode. The solution under test, which is in contact with the outer glass surface, in joined to an outer reference electrode by a suitable salt bridge. Any changes in potential between the two electrodes reflects changes in the glass electrode potential, which can then be related to alterations in ion activity within the test solution.

Satisfactory performance of reference electrodes requires that they possess :

- Reversibility
- Reproducibility
- Stability.

Reversibility is easy to achieve since modern voltmeters draw very small currents. However in case of electrodes meant for cellular applications, their small surface area can result in a very high current density. In such circumstances, the electrode should have a high enough exchange current density for the over voltage to be neglected.

Reproducibility refers not only to the ability to fabricate electrodes having identical electrochemical properties but also that responds to changes in temperature and concentration without any significant hysteresis. Stability means that electrode properties remain unchanged over long operating periods. Coupling of an electrode to the test solution via a salt bridge enhances stability by preventing their long term degradation.

7.2.1. The Hydrogen Electrode

The fact that the potential of a standard hydrogen electrode is zero at all temperatures, indicates its reproducibility. Pairs of carefully prepared hydrogen electrodes immersed in the same acid exhibits bias potentials of less than 10 μV. The hydrogen electrode consists of a platinised platinum electrode immersed in an acid solution with hydrogen gas bubbled through it. The platinisation process increases the effective surface area upto 2000 times, thereby reducing the over voltage. It also acts as a catalyser for the electron exchange process.

However, oxidising agents of various forms are reduced at the electrode surface causing the hydrogen molecules in solution adjacent to the electrode to be reduced in concentration, thereby changing the electrode potential. Other forms of impurities may also get reduced and deposit on the metal surface, reducing the surface area available for catalysis. The working life of the electrode gets reduced in the process. Also, there is need for a supply of pure hydrogen gas. Because of these reasons hydrogen electrode is rarely used for routine measurements. It is rather used as a standard against which other electrodes can be periodically checked as a means for periodically checking the pH of standard reference solutions.

7.2.2. Silver-Silver Chloride Electrodes

This type of reference electrode consists of either pure silver or platinum with a deposit of silver and a porous layer of silver chloride on the surface. The electrode is reversible to the chlorine ions. When immersed in a solution containing chlorine ions and when acting as an anode, Chlorine ions combine with silver to form AgCl ; similarly when the electrode acts as a cathode, chlorine ions pass into solution from the AgCl layer, depleting it in the process. The silver ion activity is thus maintained in the electrolyte at the electrode surface and it is inversely proportional to the chlorine ion activity. The solution in the vicinity of AgCl is saturated with Ag^+ and Cl^- ions and an equilibrium exists as indicated below :

$$\text{AgCl (solid)} \rightleftharpoons Ag^+ + Cl^- \text{ (equal concentrations)} \quad ...(7.8)$$

At equilibrium, the rate at which the dissociation occurs is exactly balanced by the rate of association of Ag^+ and Cl^- ions.

The presence of solid AgCl in a solution that already contains Cl^- ions through the dissociation of another substance does not change the dissociation rate of AgCl. Hence, the Ag^+ activity must be reduced in order to maintain the same association rate.

Thus, we define the "Solubility Product" given by :

$$K_s = (a_{Ag^+}) \cdot a_{Cl^-} \quad ...(7.9)$$

This product has a value = 1.78×10^{-10} at 25°C and is independent of Cl^- and Ag^+ activities. For pure water, $a_{Ag^+} = a_{Cl^-} = 1.33 \times 10^{-5}$ moles per liter, while for a solution for which $a_{Cl^-} = 1.0$, $a_{Ag^+} = 1.78 \times 10^{-10}$ moles/litre.

The relation $Ag \rightleftharpoons Ag^+ + e^-$ indicates the electron exchange process at the silver surface.

It shows that the potential of Ag-AgCl electrode is that of an Ag-Ag^+ electrode in which the activity of Ag^+ is determined by Eqn. (7.8)

Therefore from Nernst equation, the electrode potential

$$E = E^\circ_{Ag/Ag^+} + \frac{RT}{F} \ln (a_{Ag^+}) \quad ...(7.10)$$

which can be expressed in terms of Cl^- activity by substituting from Eqn. (7.9) :

$$E = E^\circ_{Ag/Ag^+} + \frac{RT}{F} [\ln (1.78 \times 10^{-10}) - \ln (a_{Cl^-})] \quad ...(7.11)$$

Since, the standard potential of Ag/AgCl = 0.2224 V (from standard tables) and $\frac{RT}{F} = 0.0256$, both at 25°C.

$$E = 0.2224 - 0.2256 \ln (a_{Cl^-})$$

In very concentrated solutions of KCl, the solubility of AgCl increases very much over the dilute solutions but the value of solubility product remains unaltered. However, the chloride layer should not get dissolved away. This is ensured by pre saturating the KCl solution with AgCl. The electrodes are prepared by electrodeposition of silver on platinum and then using the electrode as anode in HCl or KCl solution for electro forming an AgCl layer on the surface. The Ag-AgCl electrodes prepared in this way are probably the most reproducible, reliable and convenient reference electrodes, next only to the hydrogen electrodes. Pairs of carefully prepared electrodes, after an initial ageing, have a bias potential (potential difference between two electrodes in the same solution), of less than 50 μV and are stable to less than 10 μV for long periods of time.

7.2.3. Mercury-Mercurous Chloride (or Calomel) Electrodes

The commonly called Calomel electrode consists of an Hg_2Cl_2 layer about 0.5 cm thick on the surface of pure mercury with electrical contact to mercury being made by a platinum wire. When used as a reference electrode, either saturated 1 N or 0.1 N KCl solutions are used. The saturated solution is preferred because of its small junction potential with other less concentrated solutions.

Like AgCl, Hg_2Cl_2 is only weakly soluble in water.

At equilibrium, $$Hg_2Cl_2 \rightleftharpoons Hg_2^{+2} + Cl^-$$

with a solubility product given by :

$$K_s = (aCl^- . (aHg_2^{+2}) \approx 1 \times 10^{-8} \text{ at } 25°C$$

The presence of chlorine ions in the solution causes the Hg_2^{+2} ion activity to decrease to a value inversely proportional to the square of the Cl^- activity.

The electrode potential is determined by the equilibrium

$$2Hg \rightleftharpoons Hg_2^{+2} + 2e^-$$

and

$$E = E°_{Hg/Hg_2^{+2}} + \frac{RT}{F} \ln [aHg_2^{+2}]$$

$$= E°_{Hg/Hg_2^{+2}} + \frac{RT}{F} \ln [1 \times 10^{-8}] - 2RT \ln (a_{Cl^-})$$

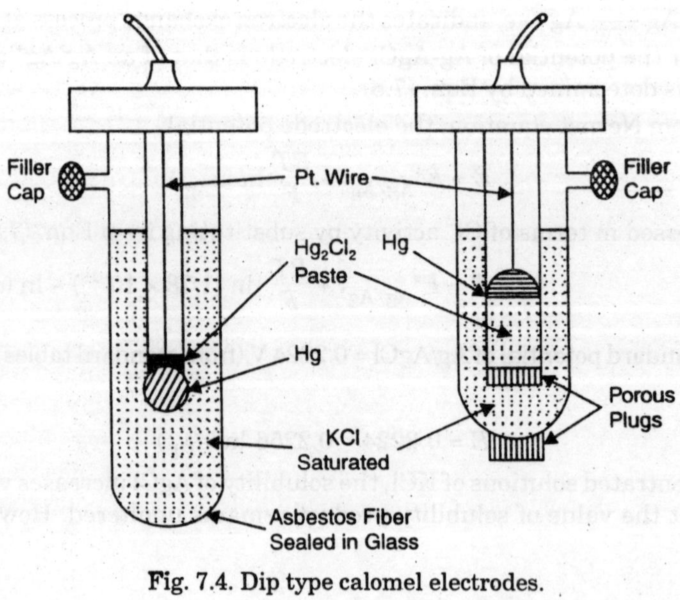

Fig. 7.4. Dip type calomel electrodes.

From standard tables,

$$E^{\circ}_{Hg/Hg_2^{+2}} = 0.797 \text{ V}$$

Therefore, Electrode potential,

$$E = 0.268 - 2RT/F \ln [a(Cl^-)]$$

or

$$E = 0.268 - 0.0512 \ln [a(Cl^-)] \quad ...(7.12)$$

Some "Dip type" calomel electrodes are shown in Fig. 7.4. In (*a*), the test solution and the saturated KCl solution are joined by the fluid flow along the asbestos fiber, while in (*b*), the liquid junction occurs at the porous plug. Though inferior to the Ag-AgCl electrode, calomel electrode is still preferred when miniaturisation of size is not important.

7.3. pH MEASUREMENT

7.3.1. Definition and Meaning

Measurement of pH has significance in analytical chemistry and clinical biochemistry. It is the measure of the degree of acidity and alkalinity of a solution. When hydronium ion activity exceeds the hydroxyl ion activity, the solution is acidic and vice versa for alkaline solution. Ultrapure water is neither acidic nor alkaline but neutral. For this hydronium and hydroxyl ion activities are equal and the reaction for equilibrium is given as :

$$2H_2O \rightleftharpoons H_3^+O + OH^-$$

Since, water is only weakly dissociated, the ion activity product

$$K_w = [a(H^+)] \,.\, [a(OH^-)]$$

If an acid is added to water, hydrogen activity increases and hydroxyl ion activity decreases by an amount equal to that required to maintain the activity product constant. Since the ion activities may vary by many orders of magnitude, the scale of measurement is conveniently compressed by using logarithm to the base 10 of the hydrogen ion activity. pH is therefore defined by :

$$pH = \log_{10} [a(H^+)] \quad ...(7.13)$$

For pure water at 25°C,

$$K_w = 1.008 \times 10^{-14} \text{ or } \sqrt{K_w} = 1.0 \times 10^{-7}$$

and since two activities are equal, $(aH^+) = 1 \times 10^{-7}$ or pH = 7.0. Therefore, pH < 7.0 indicates acidic and pH > 7.0, an alkaline solution.

The measurement of pH requires an electrochemical cell that responds linearly with pH and is sensitive to hydrogen ions only. Unfortunately, this is not practically realisable and thus to design an electrochemical cell for measuring hydrogen ion activity alone. All such measurements involve two or more ion activities. An exact calculation of an individual ion activity is not possible due to the limitations of thermodynamic methods available.

In view of this, an operational definition of pH is evolved which is based on standard buffer solutions. In the most prevalent system, a 0.05 M potassium hydrogen phthalate solution is taken as having a pH value of exactly 4.0 at 15°C. This value corresponds to the true value calculated from Eqn. (7.13) to within ± 0.01 pH units. After having defined a solution of a known pH value, the pH of all other buffer solutions can be determined. From these standard buffers, a pH meter can be calibrated and the pH of an unknown solution determined.

An electrochemical cell having a measuring electrode and a reference electrode dipped into a test solution can therefore be used in which liquid junction potentials are small. If the cell emfs for two standard buffers 1 and 2 be E_1 and E_2, then the cell emf for the test solution 'x' can be determined from the relation :

$$\frac{pH_x - pH_1}{pH_2 - pH_1} = \frac{E_x - E_1}{E_2 - E_1}.$$

7.3.2. The Hydrogen pH Electrode

As already described under 7.2.1, the hydrogen electrode for pH measurement is used primarily for checking of glass electrodes and for measuring the pH of standard buffer solutions. An added drawback is that a source of pure hydrogen is required which is not always convenient.

The potential of a hydrogen electrode is given as

$$E = \frac{RT}{F} \ln \left[a_{H^+}/\sqrt{pH_2}\right]$$

which can be rewritten in terms of pH by using Eqn. (7.13) as

$$E = \frac{-2.303\,RT}{F}\left[pH - \tfrac{1}{2}\log(760/pH_2)\right] \quad \text{...(7.14)}$$

where the potential pressure pH_2 of hydrogen is expressed in millimetres of mercury.

Thus, the potential is linearly proportional to the pH with $-2.303\,RT/F$ as the slope of the line.

7.3.3. Antimony pH Electrode

A thin layer of antimony oxide (Sb_2O_3) is normally present on the antimony surface. When immersed in an electrolyte containing hydrogen ions, the oxidation-reduction reaction is

$$Sb_2O_3 + 6H^+ + 6e^- \rightleftharpoons 2Sb + 3H_2O$$

By applying the Nernst equation to this equilibrium reaction, the electrode potential

$$E = E^\circ + \frac{RT}{F}\ln\left[a_{H^+}\right]$$

$$= E^\circ - \frac{2.303\,RT}{F}\,.\,pH$$

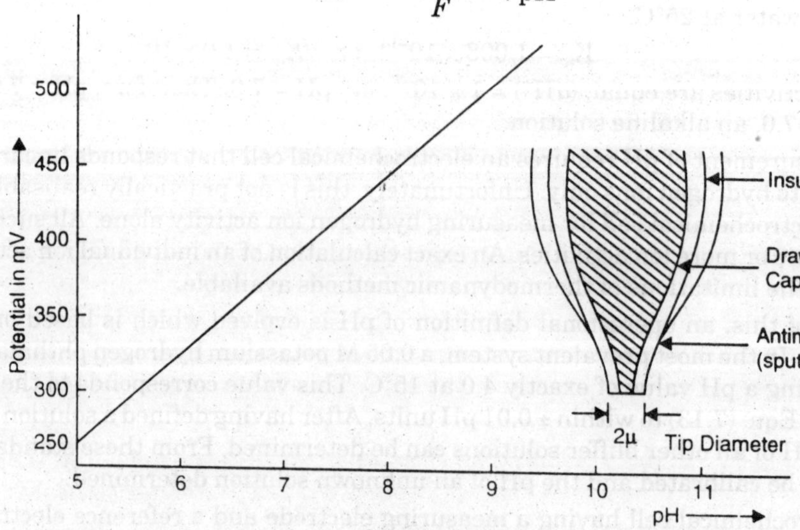

Fig. 7.5. Antimony microelectrode and its calibration curve.

where $E°$ is the standard potential of the Sb-Sb_2O_3 electrode (approx. 0.25V). It has been assumed that the activities of Sb, Sb_2O_3 and water are all unity. However, the slope of the pH characteristic is close to the ideal value of 59 mV per pH unit at 25°C.

The presence of oxidising and reducing agents significantly disturbs the potential of antimony electrodes. Still, such electrodes find useful applications in pH measurements at the cellular level. Some recent electrodes use antimony sputtered on glass in place of solid antimony. The purity of sputtering process results in good reproducibility and stability. Microelectrodes using sputtered antimony films on glass have proved useful in recording and measurement of pH within body cells and tissues. Fig. 7.5 shows the constructional details together with calibration curves and drift characteristics when put to physiological use.

7.4. GLASS ELECTRODES

When a thin glass membrane separates two electrolytes, a potential difference is observed across it. The electrode potential varies in a reproducible and well defined manner with the activities of hydrogen and other ions in the electrolytes. By a proper choice of glass composition, the potential often varies as the log of the activity of a specific ion and essentially independent of the other ion activities.

Silicate glasses consist of a network composed of silicon atoms lying at the centre of a tetrahedron formed by four oxygen atoms. The presence of an alkali metal oxide in glass causes some of its oxygen-silicon bonds to be broken. Each metal cation is bonded to a negatively charged oxygen ion with a reasonable degree of mobility. For sodium silicate glasses, the electrical conduction arises from sodium ions moving from one position to another. Since conduction is by an activated process, the resistivity will vary with temperature as exp $(-E_a/kT)$ where E_a is the activation energy.

This is usually found to be the case for most ion selective electrodes. The alkali-metal ion-bonding offers an easy possibility of exchanging the metal ion for H^+ ion. This process of ion exchange enables the glass electrode to respond to H^+ and other ions.

The penetration of water into silica glasses to form a hydrated silica layer is particularly evident with the pH sensitive glasses. The hydroscopicity and pH are therefore intimately related. The thickness of the hydrated layer varies from 10 Å to 10 μm depending on the glass composition.

Ideally, if test solution and reference solution are identical, there should be zero potential difference across the glass electrode. In practice however, an offset or an asymmetry potential remains whose value depends on the pH, prior history of the electrode and temperature.

The high resistivity of glass and consequently, the high resistance between the two glass surfaces of electrode, causes some problems in measurement of potential. Glass pH electrodes having a glass thickness of about 100 μm typically have resistances in the range of $10^7 - 10^{10}\ \Omega$ or even higher in case of glass microelectrodes.

Therefore, to avoid measurement errors arising from this high source resistance, it is essential that the potential measuring or recording instrument should have an even higher internal resistance.

With modern electrometers (solid state type), an input resistance of $> 10^{13}\ \Omega$ and an input leakage current of $< 10^{-13}$ A are achievable. This meets the demand of most glass Ph electrodes.

The problem of electrostatic interference to which all high input impedance measurements are susceptible, can be overcome by using single or double electrostatic shields.

An important aspect of glass electrode design pertains to a suitable inner reference electrode. Generally, either calomel or Ag-AgCl is electrode immersed in either dilute HCl (often 0.1 N) or a buffered chloride solution. The inner reference solution should have a buffer capacity sufficient to cope with any alkali ions leached out of glass. It should not attack glass and should have concentration necessary to ensure long term stability of the reference electrode. Unlike the external reference electrode, elimination of liquid junction potential through the use of a concentrated salt bridge solution, is not required with regard to the inner electrode.

The choice of a suitable combination of inner and outer reference electrodes is governed by the desire to achieve a small isothermal temperature coefficient for the system and to have a zero cell emf at some fixed pH value.

7.5. pH MEASURING SYSTEM USING GLASS ELECTRODES

A typical pH measuring system using a glass electrode and an external calomel electrode used as a reference electrode is shown in Fig. 7.6. There is an inner Ag-AgCl electrode immersed in a suitable buffered chloride solution, in combination with a calomel external reference electrode. By enclosing the external electrode into an outer glass sleeve, a combination unit results that is capable of measuring the pH of small sample volumes. A liquid junction of either ceramic or glass fiber is coaxial to the pH sensitive glass.

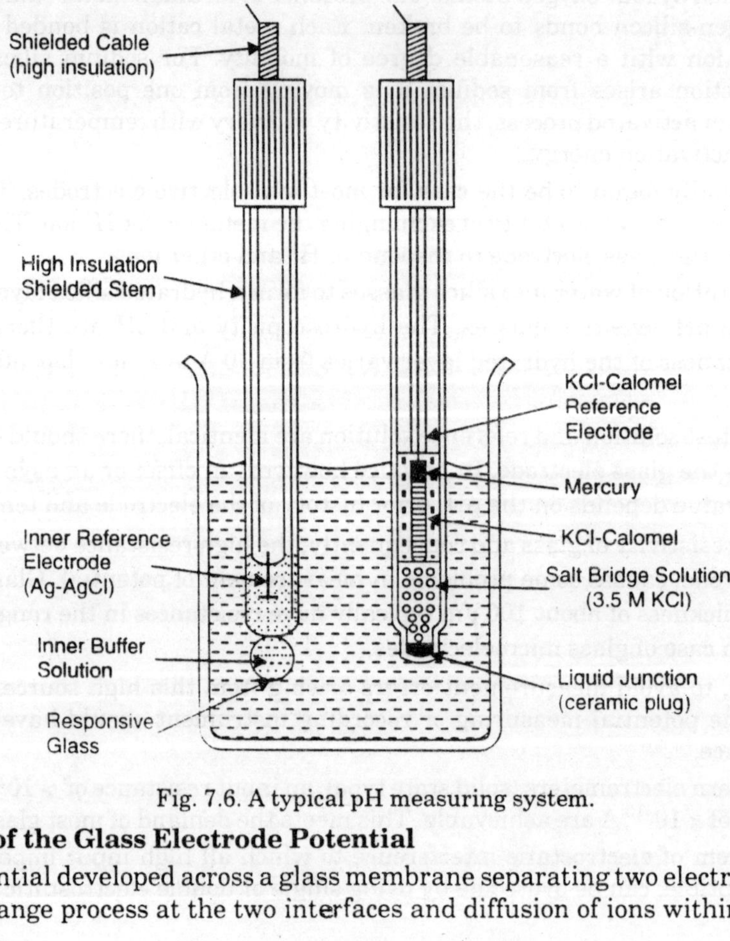

Fig. 7.6. A typical pH measuring system.

7.5.1. Theory of the Glass Electrode Potential

The potential developed across a glass membrane separating two electrolytes arises from the cation exchange process at the two interfaces and diffusion of ions within the glass. Thus, the total electrode potential can be written as

$$E = E_1 + E_2 + E_d \quad ...(7.15)$$

where E_1 denotes the phase boundary potential in solution no. 1, E_2, the phase boundary potential in solution no. 2 and E_d the diffusion potential.

Suppose ions I are exchanged for ions J in the glass and that an equilibrium exists in the form

$$I^+ \text{(glass)} + J^+ \text{(soln.)} \rightleftharpoons I^+ \text{(soln.)} + J^+ \text{(glass)}$$

The boundary potential arising from this cation exchange process is given by :

$$E_1 = \text{A constant} + RT/F \ln (a_i + ka_j) \quad ...(7.16)$$

where a_i and a_j are cation activities in solution 1

and
$$k = (a_i/a_j) \,.\, (a_j/a_i)$$

is an ion exchange constant and a' is the ion activity just inside the glass. The presence of an intruding cation in the glass with a mobility greater than the ion it replaced, results in a greater diffusion flux for this ion. Since the higher mobility cation tends to diffuse faster, a net charge imbalance will occur in glass. The resulting electric field causes the high mobility ion to be restrained to a greater degree than the low mobility ion and thereby ensures that the net flux due to each cation type are identical. Thus, the diffusion potential arises from the equilibrium requirement that there be no change in the charge distribution within the glass.

The diffusion potential can be expressed as

$$E_d = \text{A constant} + RT/F \ln \left[\frac{(a_i + k(\mu_j'/\mu_i')a_j}{a_i + ka_j} \right] \quad ...(7.17)$$

where μ' is the ion mobility in glass.

Substituting equations (7.16) and (7.17) into (7.15) and incorporating the various constants into a single term $E^\circ{}_1$,

$$E = E^\circ + \frac{RT}{F} \ln [(a_i + K(\mu_j'/\mu_i')a_j]$$

or
$$E = E^\circ + \frac{RT}{F} \ln [(a_i + K_{ij}a_j] \quad ...(7.18)$$

where the constant K_{ij} expressing the relative sensitivity of the electrode potential to the two ion types is equal to $K(\mu_j'/\mu_i')$

The more generalized form of Eqn. (7.18) could be

$$E = E^\circ + \frac{RT}{F} \ln \left(a_i + \sum_{j=1}^{n} K_{ij}a_j \right) \quad ...(7.19)$$

If ion I is H^+ and ion J is Na^+, then Eqn. (7.19) can be written as

$$E = E^\circ + \frac{RT}{F} \ln (a_{H^+} + K_H^+/Na^+ + ... a_{Na^+}) \quad ...(7.20)$$

The smaller the value of (K_H/Na a_{Na}), the closer the characteristics of the electrode will be to the ideal *E Vs* pH relationship.

The graph of eqn. (7.18) shown in Fig. 7.7 illustrates how the range of linear operation depends on the value of (K_H/Na a_{Na}).

It will be observed that 'selectivity constant' K_{ij} and the activity a_j determine the range of linearity. If the pH response is to be linear in the highly alkaline range, it is necessary that the selectivity constant be small ; otherwise, small concentrations of sodium will cause significant errors.

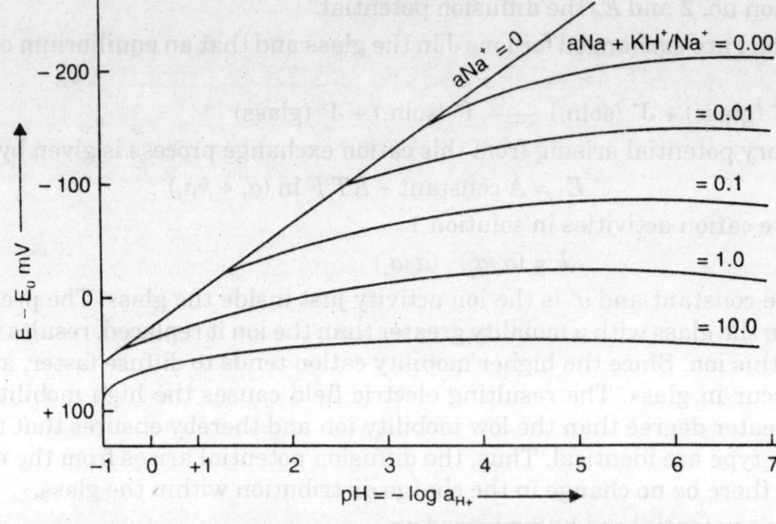

Fig. 7.7. **Graphical presentation showing the effect of selectivity and sodium activity on pH response linearity of a glass electrode.**

BIBLIOGRAPHY

1. Richard S.C. Cobbold, *Transducers for Biomedical Measurements : Principles and Applications*, John Wiley & Sons, New York, 1974.
2. R.G. Bates, *Determination of pH : Theory and Practice*, John Wiley, New York, 1964.
3. D.J.G. Ives and G.J. Jauz Eds., *Reference Electrodes : Theory and Practice*, Academic, New York 1961.
4. G. Eiseman Ed., *Glass Electrodes for Hydrogen and other cations*, Dekker New York 1967.
5. R.A. Durst, Ed., *Ion Selective Electrodes*, special Publication no. 314, NBS : Washington, D.C., 1969.
6. D. Patranabis, *Sensors and Transducers*, Wheeler, 1997.

8

Bio-Sensors And Transducers

Lately, a close interaction of biotechnology with electronics has given birth to a new class of sensors known as biosensors. Biosensors based on immobilised enzymes and micro-organisms have been exploited to a reasonable extent at international level in the areas of food and fermentation analysis, health care, environmental monitoring and even defence.

Stated precisely, biosensor is an analytical device, which uses biologically sensitive material to detect biological or chemical species directly without the need for complex processing of samples. It is made by attaching biologically sensitive materials, *e.g.* enzymes and antibodies to a suitable transducting system, which converts the biochemical response into a quantifiable and processable electrical signal. Quite logically, biosensors fall into domain of biological process engineering as well as measurement and control techniques for bio processors; high sensitivity and specificity of enzymes has led to their use as analytical tools in environmental monitoring.

Modern concepts of biosensors owe much to the ideas of L.C. Clarke Jr. (1962) and his co-workers. They proposed that enzymes could be immobilised at electrochemical detectors to form more intelligent enzyme electrode sensors (enzyme electrodes had been used as such, much earlier) which could enhance the versatility of the base detector.

8.1. CLASSIFICATION OF BIOSENSORS

Amperometric biosensors, commonly called enzyme electrodes combine the specificity or selectivity of enzymes with the analytical power of electrochemistry. The method of detection basically involves applying a constant potential between the sensing electrode and a reference electrode in the test environment and measuring the resultant steady state current.

Potentiometric biosensors involve non-faradaic electrode processes with no net current flow and upon the principle of accumulation of charge density on electrode surface, resulting in the development of a significant potential at the electrode. These sensors are mainly based on field effect transistors (FETs) and were first proposed in 1970.

Piezoelectric biosensors are principally based on the measurement of change in resonant frequency of a piezoelectric crystal as a result of mass changes on its surface. These biosensors are commonly known as surface acoustic wave (SAW) devices.

Optical biosensors have the advantages of using fiber optics or optoelectronic transducers for measuring biological reactions. Enzyme reactions alter the optical properties of some substances. Light emission from a biological element or its response to illumination may be conveniently monitored via an optical fiber or an optical wave-guide. Light is delivered to an immobilised sensing layer at the tip of the fiber and absorbance, luminescence or fluorescence could be monitored either via a second optical fiber or by the same fiber, optically bifurcated. Adoption of optical means converts chemical or biochemical concentration or activities of interest in biology, into electrical signals. These electrical signals may be compensated, calibrated and interpreted by using electronic signal processing in the instrument into a format suitable for display or for driving an actuator.

Immunosensors

These are based on the selective reaction between an analyte (antigen) and the corresponding antibody that ensures specificity of immunosensor in biological fluid analysis.

8.2. ENZYME ELECTRODES

These are also known as amperometric biosensors and based on an electrochemical process combined with an immobilised enzyme, provide an important means of measuring the concentration of certain biochemical substances.

In an enzyme electrode device by Updike and Hicks (1967), glucose concentration in a solution is measured. The enzyme used is glucose oxidase immobilised in a layer of acrylamide gel 25-50 μm thick and held in contact with the plastic membrane of a clark type oxygen electrode, as shown in Fig. 8.1. If the electrode surface is in contact with a solution containing glucose and oxygen, diffusion of both the glucose and oxygen into the enzyme layer occurs. The diffusion of oxygen through the plastic membrane of oxygen electrode is reduced in the presence of glucose owing to some of the oxygen being used in the following enzyme-catalyzed reaction:

$$\text{Glucose} + O_2 \xrightleftharpoons{\text{Glucose oxidase}} \text{Gluconic acid} + H_2O_2$$

Thus, the presence of glucose causes hydrogen peroxide to be formed and the oxygen consumed in the process causes the oxygen flux reaching the electrode to be reduced, thereby decreasing its current.

If the oxygen concentration in the solution is great enough for it not to cause any rate limitation and if the glucose concentration is not too large, it is found that the glucose concentration and the decrease in measured oxygen tension are linearly related. The response time depends on the thickness of the plastic and gel membranes. As found by Updike and Hicks for 98% response, the time taken ranged from 30 sec. to 3 min.

With the single electrode, the electrode current responds not only to the glucose concentration but depends on the oxygen tension of the test solution as well. To avoid this difficulty, a dual cathode arrangement can be adopted in which the enzyme gel of one cathode is made inactive by heating so that it responds only to oxygen. By taking the difference between the two cathode currents, an output signal is obtained that is proportional to the glucose concentration and one that is essentially independent of the oxygen tension.

In yet another alternative enzyme cathode for glucose determination, Guilbalt and Montavo (1969) have made glucose oxidase in direct contact with a platinum cathode poised at a potential sufficient to cause the hydrogen peroxide to be electro-oxidised. Either the rate of change of electrode current or the total steady state current produced can be used as a measure of glucose concentration.

In a urea-specific enzyme electrode, the enzyme urease is immobilised in polyacrylic gel. The urease catalyzes the hydrolysis of urea according to the following reaction :

$$\text{Urea} \xrightleftharpoons{\text{Urease}} NH_4^- + HCO_3^-$$

Thus, by detecting the concentration of ammonia with an electrochemical sensor, the urea concentration can be deduced. To use the electrode for estimating the quantity of urea in blood, it is necessary to use a multi electrode system to avoid the interfe-ence errors arising from the NH_4^+, K^+ and NH_4^+ ions normally present in blood.

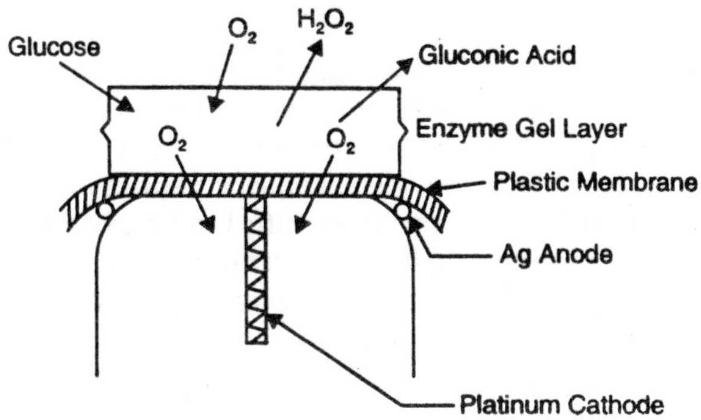

Fig. 8.1. Enzyme electrode for measuring glucose concentration.

8.3. SOME APPLICATION AREAS AND SPECIAL FEATURES

Biosensors are essentially chemical sensing electrodes modified with enzymes or microbes. These are claimed to be highly sensitive and efficient in certain areas, particularly in modern bioprocesses, where in order to achieve high efficacy in production of a desired product, it is necessary to ensure a continuous supply of substrate and monitor its concentration in- situ to overcome the problems of conversion to other undesired products.

The in situ monitoring of concentration is also necessary so as to circumvent the problem of product inhibition besides those related to frequent sample withdrawals and tedious analytical procedures currently in vogue. Moreover, biosensors are capable of sensing molecular species whereas chemical sensors sense ionic species only. Hence it is envisaged that the development of biosensors will simplify their quantitative analysis eliminating the pre-treatment step allowing in situ estimations.

Biosensors are constructed by holding a bio-catalytic substance near the tip of an ion or gas selective electrode. The electrode responds on diffusion of specific substrate for biochemical reaction into the catalytic region and is converted into by product, which the electrode can sense. The analysis is complete when potential of the electrode system reaches a steady state condition. In most cases, the enzyme is physically trapped or held at the sensing electrode surface. There have been other methods of attachment, which rely on chemical bonding of enzyme/microorganism to the surface of the membrane.

8.4. A PIEZOELECTRIC BIOSENSOR FOR ODOUR (SMELL) DETECTION

Human nose combined with a complicated olfactory system (as also in case of animals) can detect and differentiate the presence of odour (good or bad) even at trace levels of different odourous compounds. Sensory evaluation of odour is one of the important parameters in food processing, wines and beverages, air quality monitoring as well as in many cosmetic and fermentation processes/products. The odour quantification of any particular samples is normally based on conventional panel decisions based on comparative judgment. Of course the sense of comparative judgement by professionals is in turn determined by their sense of smell, taste, experience and mood. Unfortunately, there is no effective instrumental analysis to replace human sense. Analytical instruments such as GC/MS (Gas Chromatography/Mass Spectrometry), HPLC (High Pressure Liquid Chromatography) and NMR (Nuclear Magnetic

Resonance) could be used to monitor the particular compounds present in a variety of air samples ; its use is, however, impractical for monitoring unknown odourous compounds. It also suffers from tedious, time consuming and expensive works and more importantly, lack of direct correlation between the instrumental results and human perception.

A number of biochemical studies have suggested that the first step in odourant recognition involves odour molecules reacting with perceptor proteins residing on the olfactory cilia, (such studies have been conducted by various researchers very recently during 1984 to 1991) the dendrite membrane extrusions of the chemosensory neurons in olfactory epithelium. The olfactory receptor proteins (ORPs) may contain specific binding sites for particular compounds. When binding occurs at the receptor surface, the transduction of stimulus to neural signal processing eventually results in the sense of smell. Several odourant-binding proteins have been identified and isolated from the olfactory epithelium and cilia for physiological studies. Such studies have been reported till as lately as 1993 and are still continuing.

Based on the results of these studies, recently, many commercial "Electronic noses" from France, U.K., U.S.A. and Japan, under the trade names like Alpha MOS, Neutronics, odour Mapper, Mosaic etc. have become available. Their applications in a wide range of samples certainly enhance our detection of odourous molecules. The various pattern recognition techniques, *e.g.* metal oxide, and mass sensitive sensors etc. have been applied.

However, the piezoelectric crystal sensor coated with suitable chemicals, enzymes or polymers gives better results. The basic principle is that when the specific gas is sorbed onto the coated material, the increased mass will result in a reduction of resonance frequency of the crystal. Resonance frequency may also change due to viscoelastic and gravimetric effects. By monitoring the change in resonance frequency, the mass or concentration of this specific gas can be quantified by using the following relation :

$$\Delta F = -F\Delta m/(A.r.t.) \quad ...(8.1)$$

where ΔF is the change in resonance frequency (Hz), F, the initial crystal frequency (MHz), A, the total surface area (cm^2), *r*, the density of the crystal (g/cm^3), *t* the thickness of the crystal (cm) and Δm the mass sorbed (*g*).

With appropriate parameters (*e.g.* $r = 2.65$, $t = 0.168/F$, the equation (8.1) reduces to

$$\Delta F = - \times 10^6 F^2 \Delta m/A \quad ...(8.2)$$

with a total surface area A, Eqn. (8.2) is represented by the following general expression :

$$\Delta F = -K\,\Delta m \quad ...(8.3)$$

Taking into account, the amount of adsorbed gas, ΔF is proportional to the concentration C (mg/m^3) in the gas phase. Eqn. (8.3) then can be expressed as

$$\Delta F = -K_0 C \quad ...(8.4)$$

where K_0 includes constants related to piezoelectric crystal and the adsorbed gas. The assumption in Eqn. (8.4) is that the frequency change is proportional to the particular gas concentration. The sensor is able to detect the particular gas at the ppb (10^{-9}) level, SO_2 and NH_3.

The schematic diagram of the sensor with detector and analyser is shown in Fig. 8.2. Piezo crystal is 4.5 mm in dia with an initial resonance frequency of 9 MHz. Two sensor systems are used, one employes the isolated crude ORPs and the other a sensor array consisting of five separate olfactory receptors and one phospholipid coated piezo crystals, arranged side by side along a circular path on the top of a Teflon base.

The olfactory receptor proteins (ORPs) are to be isolated from bullfrogs by lengthy biochemical procedures and the same is then coated onto the surface of piezoelectric electrode, so as to closely simulate the mechanism of human olfaction. The piezo crystal serves as a signal transducer.

The Coating Method

A 1 μl aliquot of the isolated crude ORP solution is evenly spread onto each side of the crystal electrode with a blunt glass rod. The crystal is then placed in a desiccator for at least 4 hours, allowing the probe to completely dry, leaving a film of coating on the electrode surface.

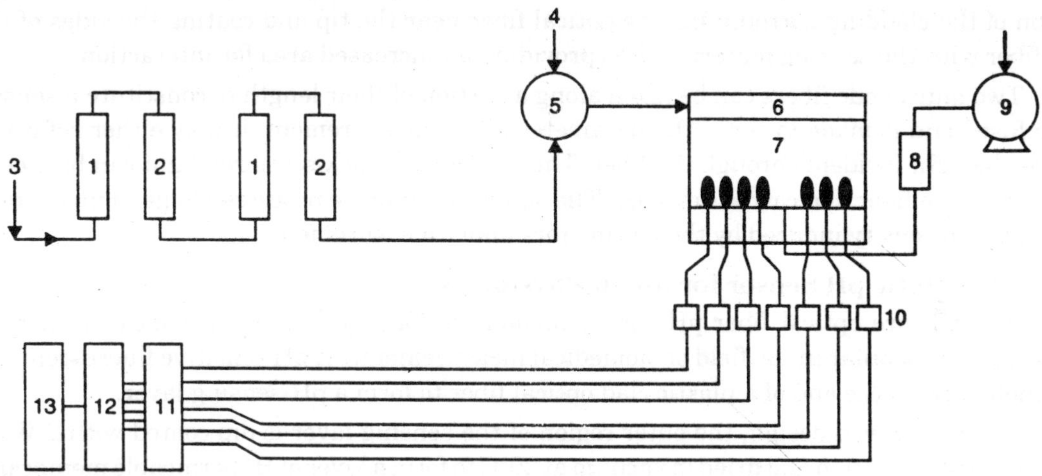

1. Silica Gel
2. Activated Carbon
3. Carrier Gas Inlet
4. Sample Inlet
5. 3-Way Valve
6. Detection Chamber
7. Multi Array Sensors
8. Flow Meter
9. Vacuum Pump
10. Oscillator Circuit
11. Frequency Counter
12. Computer
13. Monitor

Fig. 8.2. Schematic of piezoelectric olfactory biosensor system.

The amount of coating on the crystal could be determined by monitoring the initial frequency and that obtained after subjecting the piezo crystal to odourants using equation (8.3)

The ORP coated piezo crystals are referred to as odour sensing elements.

Measurements

Before any measurement of frequency is made, the detection chamber is purged with odourless dry air until the resonant frequency reaches a steady base line (F_1). The test odourant sample is then injected into the detection chamber ; it takes about 400 seconds (response time) for the resonant frequency to reach a steady state value (F_2). The detection chamber is then repurged with odourless dry air to restore the original steady base line, to be ready with next measurement. The amount of sorbed odourant (Δm) is calculated from Eqn. (8.3) by putting $\Delta F = (F_1 - F_2)$. The partition coefficient (P) of odourants to the probe film is calculated by dividing Δm by the weight of the vapour-saturated atmosphere.

8.5. FIBER OPTICS BASED BIOSENSORS

In recent years, the scope of optical techniques in instrumentation and transducers has made a quantum jump with the ready availability of low loss optical fibers and associated electro optical components. The field of fiber optic sensors has emerged as a by product of optical fiber communication technology and such sensors now find wide application in a variety of physical and chemical environments. Fiber optic biosensors are suitable for clinical applications. Their feasibility for measuring blood gases (CO_2, O_2) bound to haemoglobin has been well established. The signal obtained is proportional to the amount of biological sensing material present, thus limiting the sensitivity of miniature devices.

However, the limits of detection of the fiber optic probe may be improved by removing a portion of the cladding surrounding the optical fiber near the tip and coating the sides of the bare fiber with the sensing material, thus providing an increased area for interaction.

Two multimode fibers can be fused along a portion of their length to constitute a sensor. If the fused end is made to sense the parameter under measurement, it may either reflect or scatter the light incident through the fiber. The reflected or back-scattered light can be picked up by the same fiber and can be detected. The signal can then be processed to get a measure of modulation intensity induced by the parameters under measurement.

8.5.1. Fiber Optic pH Sensor for use in Physiology

A miniature optical fiber pH sensor membrane for physiological pH measurement is becoming very popular in the field of biomedical measurements. A pH sensitive fluorescent dye is immobilized on the end of a plastic clad optical fiber to form a pH-sensing layer.

In an improved design, the outer region of the sensing layer is dip coated with 2 WV% MPC/Methanol solution and dried in vacuum at 20°C to form a layer of H^+ permeable membrane. Another important function of the membrane is to confer the sensor with good blood compatibility. The incident light is guided by an optical fiber to excite the sensing layer. The pH dependent fluorescence emitted from the sensing layer is guided by the same fiber, reflected by a beam splitter and detected by a photomultiplier. An LS50 luminiscence spectrometer from Perkin Elmer Ltd. is used to evaluate the sensor performance.

The schematic is shown at Fig. 8.3.

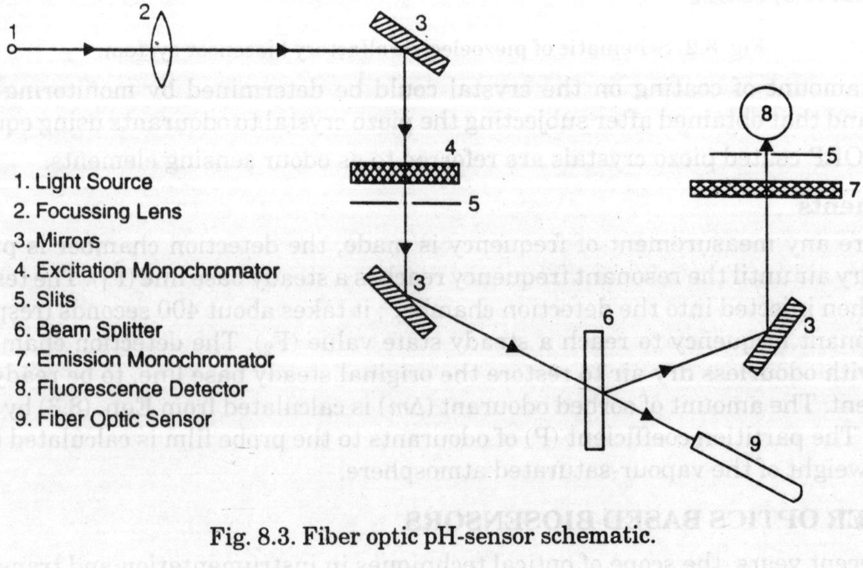

Fig. 8.3. Fiber optic pH-sensor schematic.

8.5.2. A Biosensor for Detection of Gas Toxicity

Serious discharges or leaks of volatile chemical compounds are found to cause deleterious damage to lives and also pose serious environmental problems. Some chemical compounds, e.g. volatile organic compounds, chlorinated derivatives, polyhydrocarbons etc. are very toxic to human beings and cause abnormal health effects. These toxic gaseous chemicals could be exposed to workers via air diffusion in production facility or in work places. A great variety of physical and chemical gas sensors have been developed for the detection of harmful gases but most of these lack sensitivity and selectivity in analysing the gas mixtures.

Recently, biosensors have been tried which use living organisms or biological material for detection of toxic chemicals. As a whole cell biosensor, use of genetically engineered bioluminiscent bacteria has recently flourished with applications in many areas. An advantage of using bacterial bioluminiscence is that it can be measured by the use of a fiber optic probe and light accessing equipment.

The biosensor kit (a schematic diagram shown in Fig. 8.4) in which the bioluminescent bacteria are immobilised, is attached to the end of a fiber optic probe connected to a highly sensitive luminometer to measure light output (bioluminescence) from the cells. Test chamber is a 100 ml steel cylinder with a water jacket and a thermostatic water bath, in order to maintain a constant temperature. The biosensor kit with a fiber optic probe is connected to an upside port in the test chamber by black rubber tubing so as to prevent leakage of bio-luminescence and also the entry of ambient light. After reaching a steady bioluminescence level (BL), 3 ml of benzene solution is injected into the reactor through a syringe port to give various concentrations of vapourised benzene in test chamber and BL measured every minute. To determine the real vapour concentration of benzene, 50 ml of vapour samples are taken by using tight syringes from the test chamber every minute and injected into the septa port of a gas chromatograph. An ultra 1 Capillary column and a FID detector are used for analysis.

Benzene is selected as a sample toxic gaseous chemical to show the response of toxicity to the cells. The steady state BL is maintained for about 200 minutes after the biosensor is activated by introduction into the measurement conditions. The response time is only 10 minutes, which indicates the rapid detection of gas toxicity by a biosensor. The minimum detection level of toxicity is 0.2% by volume or nearly 48 ppm of benzene in vapour phase.

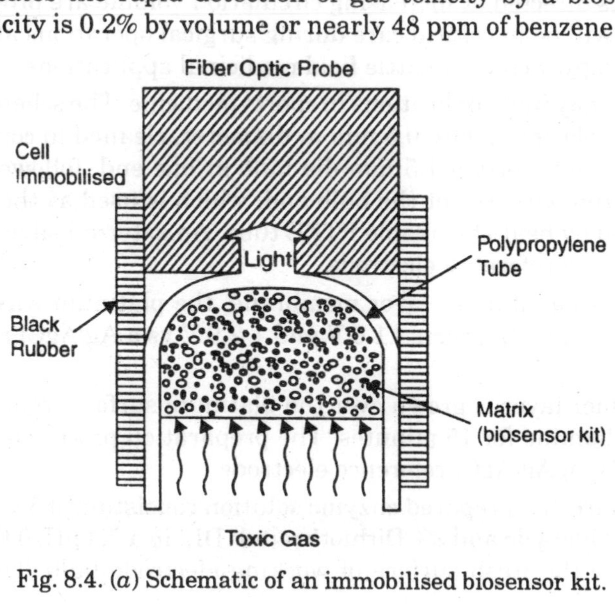

Fig. 8.4. (*a*) Schematic of an immobilised biosensor kit.

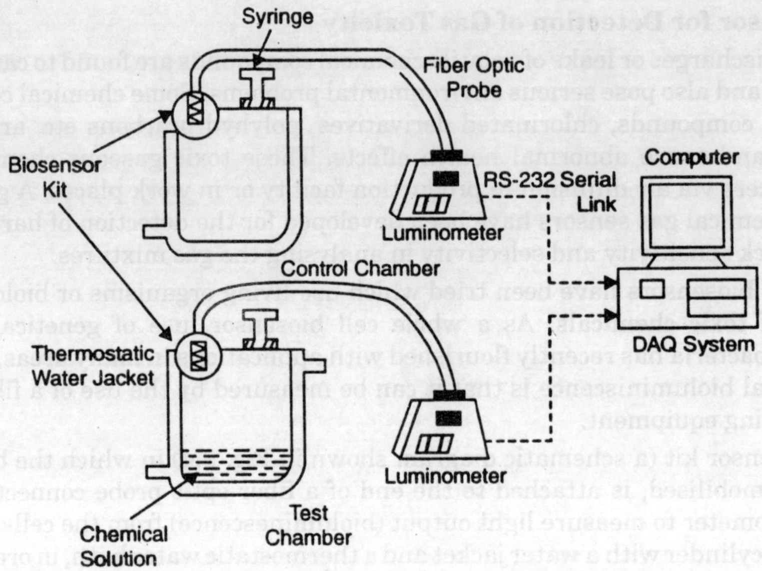

Fig. 8.4. (*b*) Gas toxicity measurement/detection system using a chamber and fiber optic probe.

The advantages of this biosensor are small size and portability.

8.6. NEEDLE TYPE LACTATE BIOSENSOR

Lactate is the key metabolite of anaerobic glycolytic pathway. Lactate concentration in physiological fluids is an indicator of energy production under anaerobic conditions of the whole organisms. Normal value for lactate level in blood is around 0.5 mM. Blood lactate values exceeding 7-8 mM are associated in some cases with a fatal outcome. The pattern of change or the trend towards an increase of blood lactate is a sensitive indicator of survival. By being able to continuously monitor the lactate in blood, the physician can get an indication of how well their rescitation efforts directed at increasing circulation volume are proceeding. The primary field of lactate sensor will be in critical care during surgical operations and intensive therapy, especially trauma therapy including battle field medicinal applications.

Lactate sensors may find applications in sports medicine. The schematic of sensor system is shown in Fig. 8.5. A platinum wire 0.5 mm in diameter, cleaned in concentrated HNO_3 and treated in propane flame to form a 1.5 mm dia bulb at one end, followed by fine sand paper treatment. The platinum wire except the bulb end, which is used as the working electrode of the sensor, is insulated by heat shrinkable PTFE tube. An external silver wire twisted around the insulation is used as a reference electrode.

Before the formation of inner layer membrane, the platinum wire is electrochemically treated by a potential cycling between – 1.0 V and + 2.0 V vs a Ag/AgCl reference electrode for an hour.

The inner polymer layer is grown over the platinum surface from a solution containing 5 mg of 1,3-phenylendiamine for 15 minutes. The preparation process is performed potentiostatically at + 0.65 V Vs an Ag/AgCl reference electrode.

A drop of 2 μl of freshly prepared enzyme solution consisting of 5% Lactic oxidase (LOD), 20% BSA, 0.5% glutaraldehyde and 2% Dithiothreitol (DL) in a 7.4 pH, 0.01 M phosphate buffer solution is delivered to the fresh surface of working electrode bulb. The enzyme mixture is

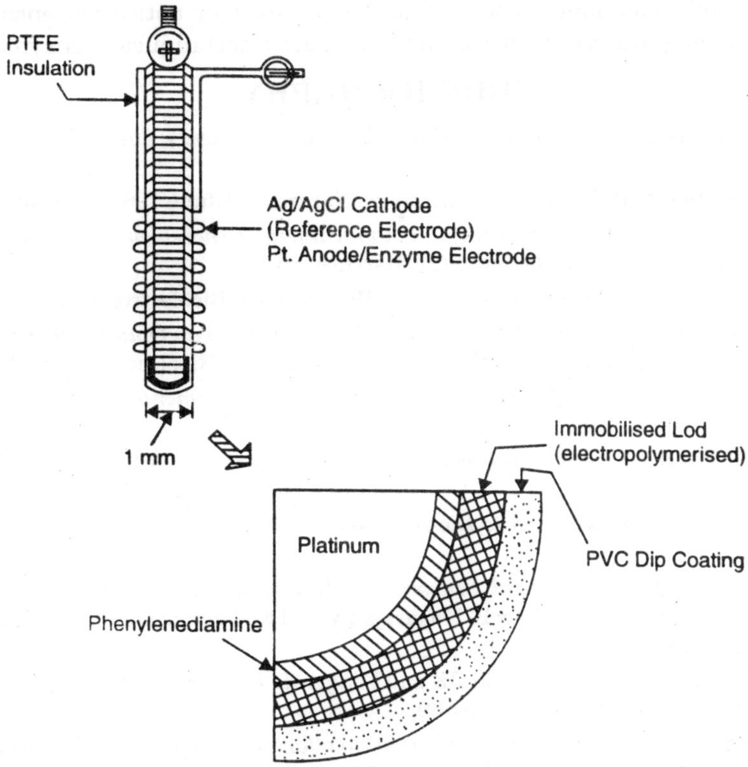

Fig. 8.5. Needle type lactate biosensor.

allowed to cross link in a moist environment for 6 hours in a refrigerator at 4°C. Electrode tip is then washed in the prepared buffer for 30 minutes. After that the sensor tip is dipped in a 12% PVC solution for 20 seconds then followed by drying in air. The electrodes are kept in moist conditions created in a refrigerator overnight.

The needle sensors are tested in vitro in a magnetically stirred, thermostatic measuring cell (50 ml vol.). A two electrode circuit is used. The catalytic working electrode is polarised as an anode and the Ag/AgCl as both reference and counter electrodes, which is polarised as cathode. A constant potential is maintained by a potentiostat. The amperometric signals are recorded.

8.7. STATUS OF BIOSENSORS IN INDIA

The R and D work in this area is still in infancy in India. Some developments have been reported from NPL, New Delhi on Lab. Scale models of glucose and Urea biosensors, which are of amperometric strip type and electrode type. The Indian Association for Cultivation of Science, Calcutta has developed a biological oxygen demand (BOD) Indicator for detecting water pollution. A similar model has been developed at CECRI, Karaikudi (T.N.) for measuring dissolved oxygen to meet the metabolic needs of anaerobic microorganisms in sewerage water. They have also developed sensors for analysing blood sugar. Centre for Biochemicals, has developed know how for isolating and purifying some key enzymes, *e.g.* Glucose oxidase, Urea and Alcohol dehydrogenase.

IIT, New Delhi has done some valuable work on fiber optics biosensors and NEERI, Nagpur has made comparative studies on immobilization techniques of enzymes.

BIBLIOGRAPHY

1. Richard SC Cobbold, *Transducers for Biomedical Measurements*, John Wiley & Sons, New York, 1974.
2. Updike SJ & Hicks GP, *The Enzyme Electrode*, Nature, 3rd June 1967, p. 986-88.
3. Anupam and R Sarin, *Potentialities and Applications of Biosensors for on line Environmental Monitoring*, IETE Technical Review, 9(2), March-April 1992.
4. R.S. Sethi, *Transducer Aspects of Biosensors*, Biosensors & Bioelectronics, 9(3) 1994, p. 243-64.
5. Zhang SF, Wickramasinghe YABD & Rolfe P, *Investigation of an Optical Fiber pH Sensor with the Membrane Based on Phospholipid Copolymer*, Biosensors & Bioelectronics 11(1-2) 1996, p. 11-16.
6. Turner APF, Karube I & Wilson G.S, *Biosensors : Fundamentals and Applications*, Oxford University Press, UK, 1987.
7. Gupta BD & Pal BP, *Fiber Optical Sensors and Instruments for Bioscience*, IETE Tech. Rev. 8(6), 1991, p. 319-22.
8. Cliffords D. Ferris, *Introduction to Bio-Instrumentation*, The Humana Press, Cliffon, N.J., USA 1978.
9. Gupta BD, *Fiber Optic Chemical Sensors and Biosensors*, Trends in Fiber optics and optical communication, Ghatak, Culshaw B, Nagarajan V and Khurana BD (Ed.) Viva Books (P) Ltd. New Delhi, 1995, p. 208-17.
10. Peterson JI, Goldstein SR, Buckhold DK and Fitzgerald RV, *Fiber Optic pH probe for Physiological Use*, Anal. Chem. Vol. 52, 1980, p. 864-69.
11. T Zong-Zeng WU, *A Piezoelectric Biosensor as an olfactory Receptor for odour detection: Electronic Nose*, Biosensors & Bioelectronics Vol. 14, 1999, p. 9-18.
12. Gil GC, Mitchel RJ et. el., *A Biosensor for Detection of Gas Toxicity using a Recombinant Bioluminiscent Bacterium*, Biosensors & Bioelectronics, 15(1-2), March, 2000, p. 23-30.
13. Yang Q, Atanasov & Wilkins E, *Needle Type Lactate Biosensor*, Biosensors & Bioelectronics, 14(2), Feb. 1999.
14. Neufeld T, Eshkenazi Imma Eshkenazi et. el., *A microflow biosensor for organophosphorus Pesticide*, Biosensors & Bioelectronics, 15(5-6) Aug., 2000 p. 323.
15. Joshi MG, *Application of Fiber optics to Biosensors : A Review*, Proc. National Conf. on Sensors & Transducers, CSIO Chandigarh Sept. 20, 21, 1996, p. 172-78.
16. Khan SA, *Supplement on Sensing Devices*, Communications in Instrumentation, 3(1) 1995, p. 46-51 and 3(4) 1995, p. 193-200.
17. Guilbault GG, *Analytical uses of Immobilized Enzymes*, Dekker, New York, 1984.
18. Christian GD, *Electrochemical methods for analysis of enzyme systems*, Levine SN(Ed) Advances in Biomedical Engineering and Medical Physics, Vol. IV, Wiley New York 1971 p. 95-161.

9

Optical And Fiber Optic Transducers

Optical sensing techniques are invariably employed for detecting radiation and sensing temperatures of bodies emitting radiation by non contact means. The technique is very often called Infra red (IR) pyrometry or radiation pyrometry, which is the only dependable means for measuring very high temperatures right upto 6000-7000 °C. Thermocouples of all varieties rarely go beyond the range of 1750 °C and resistance thermometers or thermistors are limited upto a still lower range. Fiber optics sensors are the latest in the line where optical fibers are used as light guides or wave guides for a variety of measurements. So this chapter is confined to conventional optical techniques adopted for radiation detection and non contact radiation pyrometry and novel features as well as astonishing capabilities of fiber optics when used as sensors and transducers.

9.1. RADIATION AND ITS MEASUREMENTS

Let us consider a body at a temperature T. It receives a certain amount of radiation from the surrounding environment at a wavelength λ. A fraction of it is transmitted as energy at wavelength λ, a part is reflected and a part is radiated. For equilibrium, the energy absorbed by the body at any wavelength must be equal to that emitted by it at the same wavelength, or

$$E(\lambda) = a(\lambda)$$

We select wavelength λ at values where the body appears opaque (hence transmission is negligible). Such materials as glass, plastics etc. are opaque in selected IR bands. A body that absorbs all incident energy (an ideal case), is referred to as "Black Body". A cavity in a hollow body (provided the size of cavity is small compared to the internal dimensions of the body), with all parts being at the same temperature, is a close identification of a perfect, ideal black body. Such a body will absorb all incident radiation without reflecting any and give out radiated energy at all wavelengths. Fig. 9.1 shows a plot of spectral radiant intensity or emitted power from the black body in W/cm^2. mm against wavelength k in mm, at different black body temperatures. From this, it can be observed that :

- At every temperature of black body, there is a definite amount of radiation at every frequency.
- The radiation is zero for $\lambda = 0$ and also for $\lambda = \infty$. At any black body temperature, the radiation increases with λ, reaches a peak and drops back to zero as $\lambda \to \infty$.
- The value of λ, for which W_λ is a maximum, reduces steadily as the black body temperature rises. In turn, the body temperature changes from dull red to white as the black body temperature rises. In fact the wavelength λ_p corresponding to $W_{\lambda max}$ is empirically given by :

$$\lambda_p = 2891/T \ \mu m$$

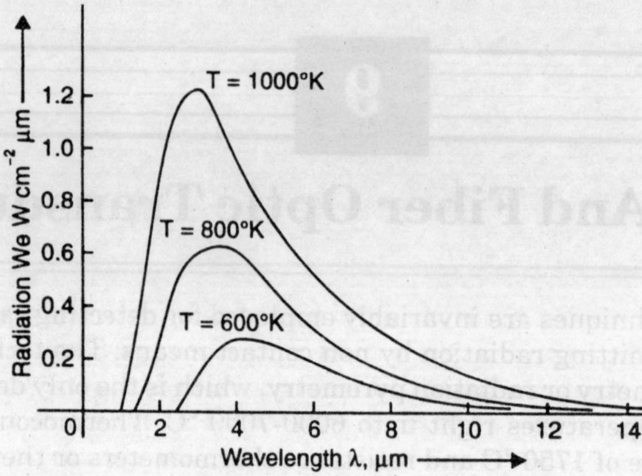

Fig. 9.1. Variation of black body radiation with wavelength.

The radiation emitted from a black body is thus a function of spectral wavelength λ, given by :

$$W_\lambda = \frac{c_1 \lambda^{-5}}{\exp (c_2/\lambda T)^{-1}}$$

where λ is in metres

and T, absolute temperature of the black body in degrees Kelvin. The values of constant C_1 and C_2 are given by :

$$C_1 = 37143 \text{ W}\mu\text{m}^4/\text{cm}^2$$

$$C_2 = 14388 \ \mu\text{mK}$$

This is called Planck's equation ; this when integrated w.r.t. λ from 0 to ∞, yields :

$$W_t = 5.67 \times 10^{-12} \text{ T}^4 \text{ W/cm}^2$$

where W_t is the total radiation from 1 cm^2 of the body surface at temperature T.

This is called Wein's equation.

The relative amount of radiation from a body at a wavelength λ is called emittance ε_λ, which is a function of λ and T. The bodies which have a constant emittance at all wavelengths, are called "Grey bodies" and the Wein's equation for grey bodies is

$$W_t = 5.67 \times 10^{-12} \ \varepsilon_r \text{T}^4 \text{ W/cm}^2$$

where ε_r is the emissivity of the body.

This equation is also sometimes known as Stephen. Boltzman law stated as :

$$W_t = \sigma T^4,$$

where $\sigma = 5.67 \times 10^{-12} \ \varepsilon_r$.

And gives the value of spectral intensity of radiation, in terms of black body or grey body temperature T. In case of two bodies at temperatures T_1 and T_2 ($T_1 > T_2$), radiation energy transfer takes place in accordance with the relation :

$$W_{t1,2} = \sigma(T_1{}^4 - T_2{}^4)$$

In practice, any black body can be taken as grey, if its emittance is sensibly constant over the significant 3-20 μm wavelength band in the IR region. This range becomes smaller at higher temperatures. This is the reason why IR pyrometry is convenient and more accurate at higher temperatures.

9.1.1. Emissivity

Emissivity characteristics of the target and the absorption in the intervening media, are very significant to the IR pyrometry. An opening of a hollow sphere, with a uniform temperature in its inside surface, is close to an ideal black body and it has an emissivity value $\varepsilon_r = 0.998$ (ideal value unity). Its value is determined by the molecular structure and the surface smoothness characteristics. In general, polished surfaces have a low emissivity. It increases and approaches unity at smaller values of λ. For most opaque bodies, it is in the range 0.85 to 0.90. For most unoxidised metals ε_r is in the range 0.2 to 0.5. For gold, silver and aluminium, it is between 0.02 to 0.04. IR pyrometry is difficult to use with them.

When a target is viewed to sense IR radiation emitting from it, the absorption and transmission characteristics of the medium between the target and the IR sensor, affects the results. Some schemes using sensor at two frequencies for which the absorption and transmission are the same and obtain temperature from their ratio. Others sense radiation in a select band of wavelengths over which the transmittance is sufficiently high. Thus if there is an intervening air column, the 8-14 μm wavelength band is used. The absorption in this band is relatively low and hence the transmittance is high.

9.1.2. Methods of Sensing Temperatures

A. Indirect Detection Method

Here, the radiation from the body is made to fall on an isolated thermal mass of very low reflectivity. The entire received radiation is absorbed by the thermal mass and its temperature rises which is taken as the source temperature. This mode of operation is in accordance with the Wein's equation. However, the active range of operation becomes limited by the non-linear relationship between the energy received and the source temperature (W_t is proportional to the fourth power of source temperature).

As shown in Fig. 9.2, the temperature sensed will differ from the casing to the sensor end and therefore losses from the detector are not precluded. This is known as a total radiation pyrometer.

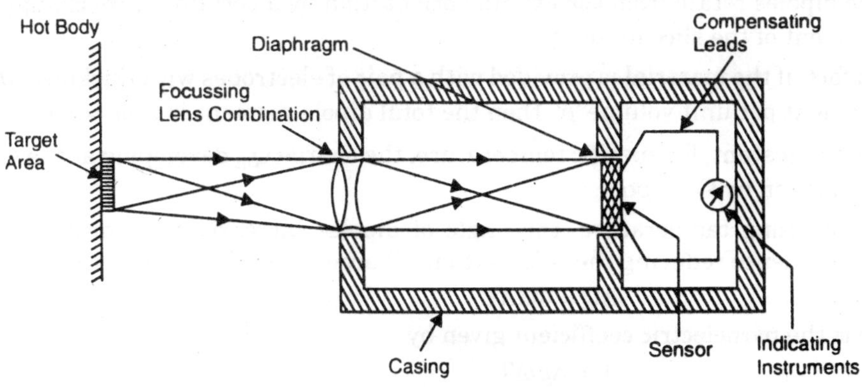

Fig. 9.2. Total radiation pyrometer.

If the detector temperature finally settles at a temperature T_2 with the casing tempeture at T_0 and target (source) temperature at T_1, then the approximate heat loss becomes

$$K_0(T_2 - T_0) = K_1T_1^4 = \text{Heat input}$$

The temperature T_0 is the cold junction temperature in case, thermopile or thermocouple is used as sensor. Then the output voltage is proportional to $(T_2 - T_0)$ which is in turn, proportional to T_1^4. However, the empirical relation, $V = K_v . T_1^n$ is more acceptable ; the value of n could be anywhere between 3.5 to 4.5, and value of constant K_v could be obtained through experimental calibration.

Among the radiation sensors used, thermopile is easily adaptable to industrial applications, it has quite low response time, approximately 0.1 sec and gives a larger output. All the hot junctions are on the thermal sink that receives radiation. Cold junctions are on a separate heat sink. J & K type thermocouples generally constitute thermopile.

Thermistors or semiconductor bolometers are less commonly used because of their poor repeatability inspite of having quite low response time and high sensitivity. It also covers only a very limited range of temperature.

Semiconductor temperature sensor with a sensitivity 1 μA/°C has come into use in recent years. It has a linear output throughout the range.

Photoelectric cells have fastest response speed and high sensitivity but again their use is limited due to their limited spectral sensitivity.

Pyroelectric detectors can be used to sense a temperature change. These are electroceramics and some of these get polarised on heating, the extent to which these get polarised, is a measure of the temperature change. Pyroelectric materials form the dielectric or a sensor capacitor. The increase in polarization gets reflected as an increase in the capacitance value. This is converted into an electrical signal using capacitance transduction techniques.

The molecules in a pyroelectric material have a permanent electric dipole and these molecules are randomly oriented in the normal ambient conditions. Above a certain temperature called Curie point, these molecules acquire rotational mobility and a piece of pyroelectric material, if subjected to an electric field, would have its molecules aligning parallel to the field. Once this operation is performed, orientation of molecules does not change even if temperature is brought down below Curie point and electric field removed. Without applying any field now if the temperature of ceramic is increased by allowing thermal radiation to fall on it upto Curie point, the dipoles rotate from the existing orientation by a certain angle, changing the overall dipole moment of the slice material.

In fact, if the material is provided with a pair of electrodes with thickness 'd', area 'α' and dipole moment per unit volume 'p', then the total dipole moment is : $m_t = pad$.

At any instant, for specific temperature, the charge Q_m developing at the surfaces of this material is given by $Q_m = p\alpha$.

With rising temperature, the angle of dipoles increases from orientation parallel to electric field, hence reducing the polarisation. Charge 'Q' on electrodes will be now in excess by

$$\Delta Q = \Phi\alpha\Delta T$$

where Φ is the pyroelectric coefficient given by

$$\Phi = \Delta p/\Delta T$$

and the corresponding voltage

$$\Delta V = \Phi\alpha\Delta T/C$$

[C being the capacitance formed with electrodes]

The dipole electrode is illustrated in Fig. 9.3 while the calibration curve of a radiation pyrometer is shown in Fig. 9.4.

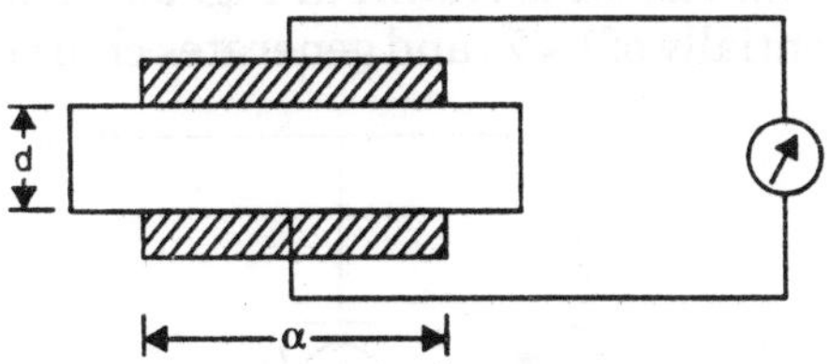

Fig. 9.3. Dipole electrode.

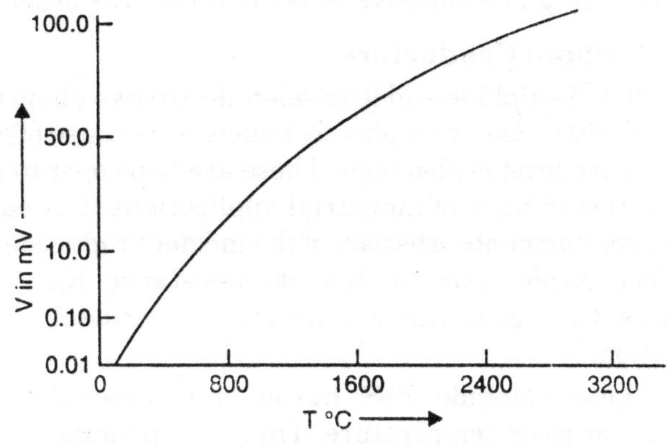

Fig. 9.4. Calibration curve of radiation pyrometer.

B. Direct Detection Method

In direct methods, the intensity of radiation is measured at one or more wavelengths. The sensors used here have to be more sensitive for this purpose or otherwise, we have to compromise on resolution. In general, a radiation sensor has a crystal lattice structure with a low energy gap between the valence and conduction bands. Part of the incident radiation on it causes the charge carriers to be liberated, which causes an electrical parameter, *i.e.* resistance, current or voltage associated with the sensor and its circuit, to change. This change is related to the total radiation received. For radiation at wavelength λ,

$$E = hc/\lambda q$$

where, E = Energy in electron volts, h = Planck's constant

c = Velocity of light, q = Charge on the carrier

For a material of energy gap E_g, incidents photons of wavelength < λ_c are absorbed and cause charge carriers to be released.

Here, $\lambda_c = hc/qE_g = (1.238/E_g)\mu m$, E_g is in eV.

Part of incident radiation is reflected and rest absorbed. With increasing depth, more and more is absorbed and converted into charge carriers. The electron-hole generation rate is therefore

$$G(x) = (1 - r)E_q a e^{-ax}$$

where, r = Fresnel reflectance, E_q = Photon irradiance and

a = Absorption coefficient

The total charge carriers N generated in a depth d

$$N = \int_0^d g(x)\,dx \quad \text{or} \quad N = (1 - r)E_q[1 - e^{-ad}]$$

The ratio N/E_q represents the number of electrons produced per photon of irradiant energy. This ratio is termed "Quantum efficiency" of the detector material η or

$$\eta = (1 - r)[1 - e^{-ad}]$$

The absorption coefficient 'a' is a function of incident radiation wavelength and is low for $\lambda > \lambda_c$ and increases steeply for different detectors. It is about 30% for extrinsic photoconductors and 10% for photo emissive sensors. It is of the order of 1% only, for photographic films.

9.1.2.1. Photo Conductors

Lead sulphide and lead selenide are two photo conductors. Both are of intrinsic type and have η = 60%. Extrinsic photoconductors, because of sparse presence of dopants, have η = 30%. Their noise level is also high. These are to be operated at cryogenic temperatures and hence of no interest in most of industrial applications. The equivalent circuit is shown in Fig. 9.5. The photo-conductor absorbs part of the incident radiation, essentially of $\lambda < \lambda_c$ and generates charge carriers inside. Due to this, its resistance R_D reduces. Change in this is a measure of incident radiation.

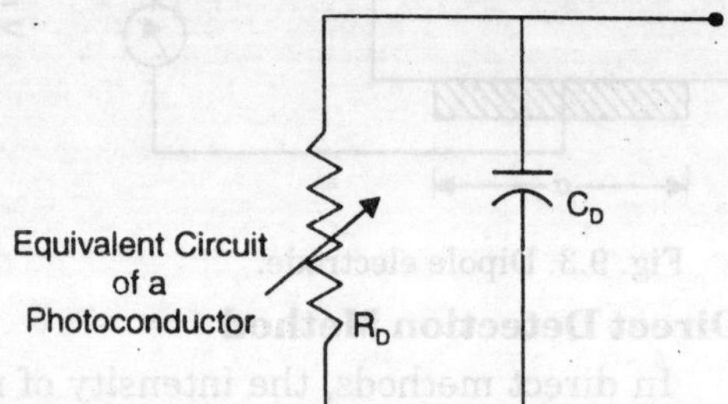

Fig. 9.5. Equivalent circuit of a photoconductor.

Lead Sulphide (PbS) has an energy gap of 0.37 eV at room temperature. The corresponding λ_c = 3.3 μm. It can be used at a radiation detector upto this wavelength. The operating temperature can be upto 80°C. Above this temperature, oxidation renders it ineffective as a sensor. Lead Selenide (PbSe) has its spectral response from 1.5 μm to 5.8 μm at room temperature. Both PbS and PbSe have linear response over a very long period at room temperature.

9.1.2.2. Photo Voltaic Sensor

These are *p-n* junctions exposed to radiation. Some electrons from the n region of *p-n* junction migrate to *p* region and holes move from *p* to *n*. This creates a potential barrier across the junction and either side is depleted of the charge carriers over a restricted zone. Subsequently, there is no net charge transfer across the junction and therefore no current through it. Any radiation incident at or near the junction, therefore gets absorbed in the restricted zone and results in electron-hole pairs. A voltage thus gets developed across the junction or a current flows through it depending on the circuit conditions. The incident radiation on the material away from the junction may also cause generation of electron hole pairs. They recombine within a short time and distance and do not contribute to any net output current or voltage. The current and voltage of a diode are related as

$$i = i_0 \left[\exp \frac{(qv)}{\text{bKT}} - 1 \right]$$

where i_0 = Reverse saturation current, v = Applied voltage
q = Charge on the electron, b = Non-ideality factor
K = Boltzmann constant and T = Temperature in degrees K.

$i = 0$ for $v = 0$. When radiation is incident on the diode, it causes an additional current i_g to flow then

$$i_g = \eta \Phi_d q \text{ and } \Phi_d = \text{A}E_q$$

where Φ_d = Total incident photon flux
A = Total active detector area and
E_q = Irradiance

With this, the total diode current :

$$i = i_0 \left[\exp \frac{(qv)}{bKT} - 1 \right] - \eta \Phi_d q$$

Silicon photodiode is the most common one. Its spectral response is within 0.2 – 1.15 μm with peak at 0.9 μm. It is usable within visible and near IR regions. Germanium photo diodes have spectral response range of 0.6 – 1.8 μm, thus covering a wider IR range. They offer good sensitivity at a low cost.

9.1.3. Direct Method : Single Colour Pyrometer

This type is shown in Fig. 9.6 and makes use of a heated filament in the field of view (FOV). The filament current is adjusted such that the filament disappears in the background. The use of red Cadmium filter makes the radiation appear monochromatic to the eye. The wavelength of such a filter is a band of 1 μm at 65 μm. The filament current indication is an indication of temperature and the milliammeter is calibrated in terms of temperature. The method essentially matches the intensity of radiation from filament to that of the source in the FOV, at a selected narrow bandwidth. The internal radiation reference source can also be an LED and its current can be adjusted like the filament current. The colour comparison is done by directing the radiation from either source alternately on to a sensitive detector, usually a photo multiplier tube. LED current is adjusted until the two currents match. The LED current is a measure of target temperature. The scheme is simple and less costly. However, it requires the ε_r of the target to be known. Normally short wavelengths (Red : 0.64 – 0.65 μm) are used so that the sensitivity to changes in ε_r is low.

The disappearing filament type optical pyrometer was the only reliable method earlier but it is giving way gradually to a two-colour comparison method, in view of electronic circuits becoming versatile and economic.

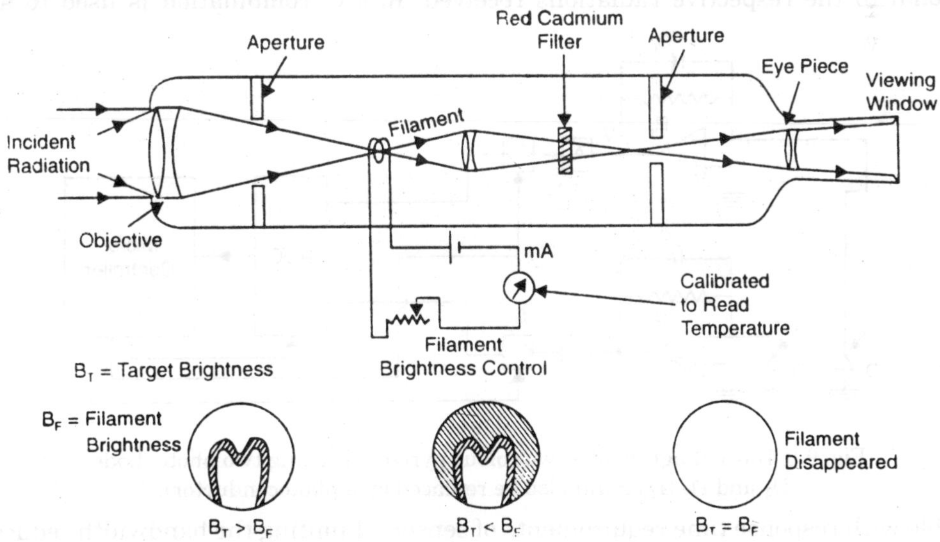

Fig. 9.6. Single colour pyrometer (disappearing filament type).

9.1.4. Two-Colour Pyrometer

This method compares the intensity of radiation at two wavelengths λ_1 and λ_2. Once the wavelengths are known, temperature T is directly related to the ratio of the respective intensities. In practice, total radiation over a narrow window of $\Delta\lambda$ is received in each case and compared. Emissivity of the material does not affect the reading and since only the relative intensities at two wavelengths are compared, the object need not fill the FOV. If λ_1, λ_2 are so chosen that the emissivities of the intervening gases, moisture glass window etc. are the same, the final reading does remain unaffected.

Detectors are specifically tailored for this method. These make use of a sandwich structure with silicon photo-diode of transmitting type on the top and a Germanium photodiode or a photoconductor (PbS or PbSe) type, underneath. The top Si photo diode sensens the radiation at 0.94 μm. The second wavelength λ_2 is chosen depending on the material used : 2.2 μm with PbS, 3.8 μm with PbSe or 1.55 μm with Ge. Both λ_1, λ_2 are chosen to be lower than λ_p, the wavelength at which the radiation is maximum. This sets an upper limit to the temperature to be sensed. Ideally, it occurs at the temperature for which $\lambda_2 = \lambda_p$, the practical limit is somewhat lower. (refer to Fig. 9.7). With this constraint, Si or PbSe sensors can be used upto 450°C, PbS and Si upto 1000°C and Ge and Si, upto 1500°C.

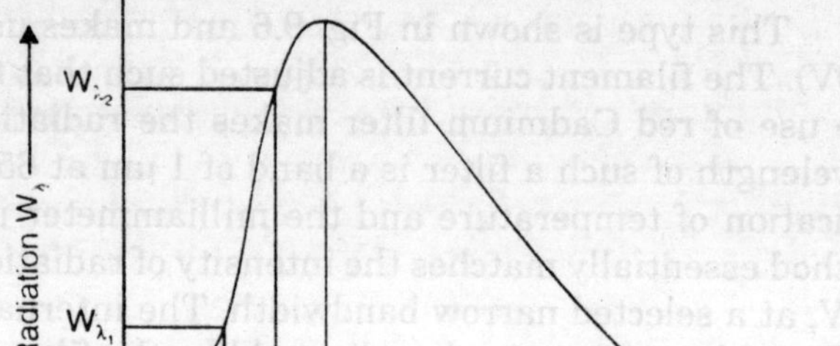

Fig. 9.7. Wavelength selection for a two-colour pyrometer.

A typical arrangement is shown at Fig. 9.8. The two detectors are D_1 = a photo voltaic diode and D_2 = a photo voltaic diode or a photo conductor. The two voltages V_1, V_2 are proportional to the respective radiations received. $R_f \parallel C_f$ combination is used to a cut-off

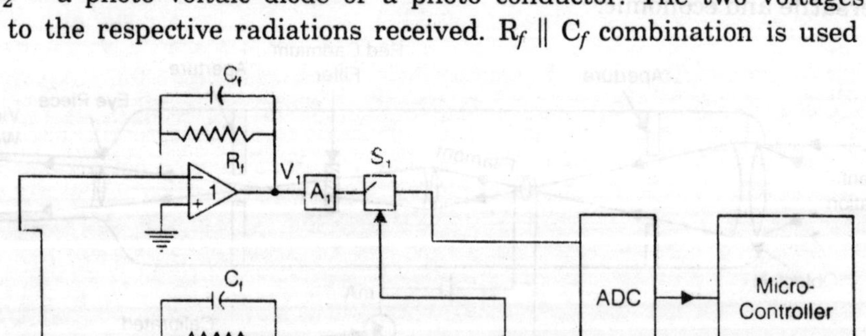

Fig. 9.8. Block diagram of a two colour pyrometer using two photo diodes D_1 and D_2 (D_2 could also be replaced by a photoconductor).

compatible with response time requirements of sensors. Limiting the bandwidth reduces noise at the operational amplifier. Amplifiers A_1, A_2 amplify and band limit these signals. The gains are adjusted such that the full scale output levels correspond to the full scale input levels of the ADCs. The micro-controller computes the ratio of the received radiations and computes the source temperature according to the equation

$$\frac{W_{e1}}{W_{e2}} = (\lambda_1/\lambda_2)^{-5} \exp\left[-\frac{C_2}{T}\left(\frac{1}{\lambda_1}-\frac{1}{\lambda_2}\right)\right]$$

which is obtained by dividing the Wein's equations corresponding to λ_1 and λ_2, viz.

$$W_e = c_1\lambda^{-5} \exp(-c_2/\lambda T)$$

The two colour method avoids inaccuracies due to absorption in the intervening medium. The numerical value of ε_r of the target need not be known. The only constraint is that ε_r of the target need not be known. The only constraint is that ε_r at both the wavelengths should be the same. Thus, the temperatures of different targets can be measured using the same instrument without recalibration or readjustment. Also, unlike the total radiation detection method, the target need not fill the viewing area.

Two colour sensors suited to individual applications are available, sensing radiation at specific wavelengths in the bands suitable for the target concerned. 7.9 μm wavelength is suitable for plastics upto 600°C, 4.8-5.2 μm for glass and ceramic surfaces in 300-1500°C range, 3.86 μm for monitoring flames in 500°C- 1500°C range, 1.1-1.65 μm is used for viewing molten metals through glass with a wide range of 250°-2500°C. The corresponding instruments use two separate sensors with the received radiation directed to either of them, through optical filters. The radiation is directed to them through mechanical choppers.

9.2. FIBER OPTIC SENSORS

9.2.1. Basic Principles and Design

An optical fiber is the most popularly used optical waveguide. It is basically a thin glass wire with overall thickness of about 100 μm almost the same as that of human hair. The light radiation given at one end of fiber travels along to the other end by means of a series of total internal reflections from the sides. This guiding action is maintained even if the fiber is bent quite sharply, so that the fiber can be handled in much the same way as copper wire. The ease with which optical fiber sensors can be interfaced with optical telemetry links is just one of the several advantages which they offer. Optical fibers are dielectric and electrically passive elements. They require no electrical power supply at the point of measurement, neither do they provide electrically conducting paths. They can therefore, provide sensors that are intrinsically safe in hazardous environments. Optical fiber sensors are also simple and compact as these usually have no moving parts, their installation is straight forward as cost is low.

In order to understand the ways in which optical fibers may be used as sensors, it is necessary to know how light radiation propagates through a fiber.

Cross sectional structure of an optical fiber is shown in Fig. 9.9. Central core is normally of glass or silica, surrounded by a cladding also of the same material but having a slightly lower refractive index. This facilitates total internal reflection of the light ray striking the core/cladding interface for all angles exceeding θ_c (the critical angle). Dimensions of fiber are comparable to the wavelength of light so that if a reasonable monochromatic source of light is used, recognizable interference patterns are formed within the fiber core through the interaction of many bouncing waves. These interference patterns are called "modes" of propagation and these depend upon the geometry and wavelength of the light used ; the distribution of light power amongst the allowed modes will depend on the initial conditions for launching of light, and on any geometrical or other perturbations which are capable of transferring power from one mode to the other.

However, when a ray of light suffers reflections at the boundary between the two media, its state of polarization also gets altered. Because, the electric field component in the plane of incidence will experience a different phase shift from that of the component perpendicular to it. This renders the polarization behaviour of light through the fiber with a large number of allowable modes, rather complex and therefore difficult to be used in a measurement system. It is of course possible to arrange the geometry of the fiber, in relation to the wavelength of light used, so that only one mode is allowed, resulting in only one angle of total internal reflection. This case [Fig. 9.9 (*c*)], known as that of single mode propagation, ensures that the light exists in a single definable state of polarization at each point along the length of the fiber. However, one practical disadvantage of a single mode fiber is that only a laser source can successfully launch light into it and laser sources are relatively expensive.

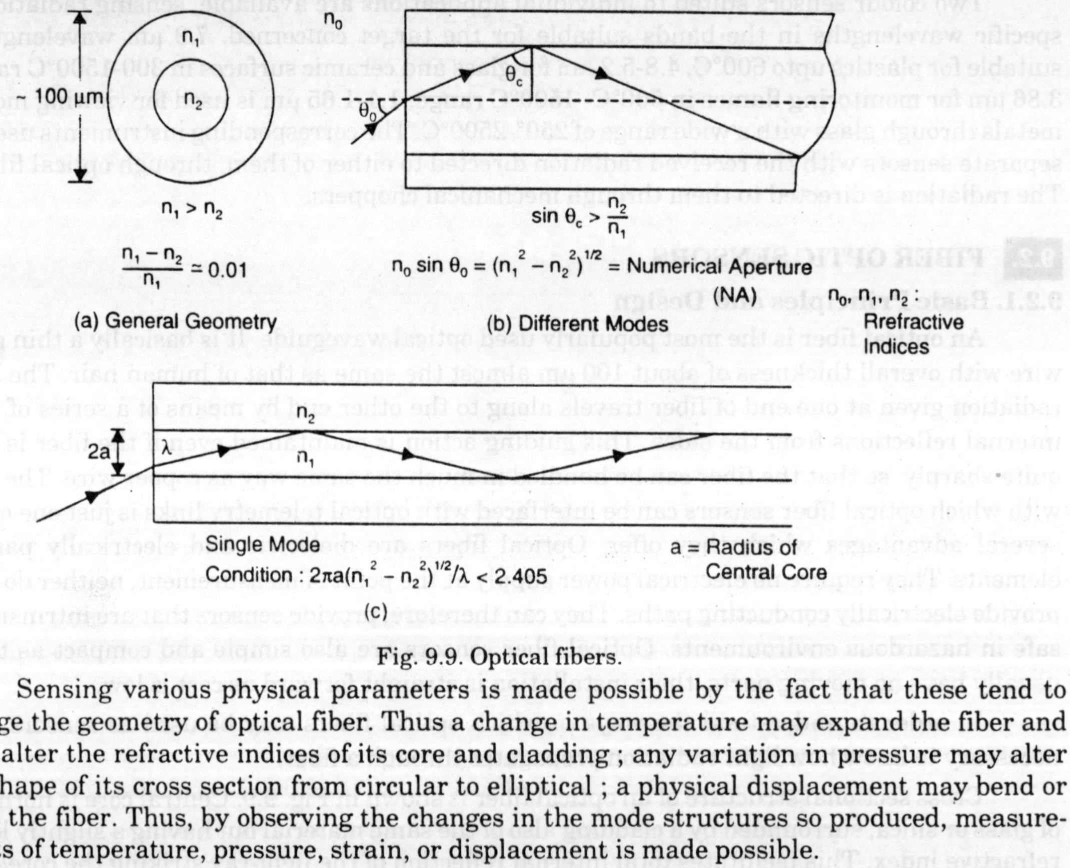

Fig. 9.9. Optical fibers.

Sensing various physical parameters is made possible by the fact that these tend to change the geometry of optical fiber. Thus a change in temperature may expand the fiber and thus alter the refractive indices of its core and cladding ; any variation in pressure may alter the shape of its cross section from circular to elliptical ; a physical displacement may bend or twist the fiber. Thus, by observing the changes in the mode structures so produced, measurements of temperature, pressure, strain, or displacement is made possible.

Also, when light passes through solid materials, it interacts with its atomic arrangement. Application of external electric or magnetic fields tends to restrict the directional movements of electrons constituting the atoms. Therefore, these external fields will also affect any light that is propagating through the material and affect its state of polarization as well (since the polarization state is determined by the structural directionality). The optical fiber has to be made with carefully controlled and predefined properties so that external influences, if any, (other than the parameters to be sensed or measured) may not affect at the same time. Also, to avoid any sensitivity to external influences, one must ensure that the optical detection system is sensitive to received light power only and not to the mode structure through the fiber.

Optical fibers offer several advantages as sensors, such as light weight, non-obtrusive, all-passive configurations, low power consumption, immunity to EMI/RFI, high sensitivity and large bandwidth, geometric versatility, flexibility, compatibility with optical fiber data transmission and processing long life with low cost, unique feature of distributed sensing and multiplexing of sensor signals.

Different types of fiber optic sensors are briefly discussed below.

9.2.2. Fiber Optic Strain Sensor

This makes use of the sensitivity of mode structure to optical fiber bending. If the fiber is bent sharply (when stressed), the angle at which some rays strike the core-cladding boundary will be so small that total internal reflection will not take place and these rays will consequently be lost alongwith the respective modes. At a certain bend radius determined by the fiber's geometry, the bend loss will become steeply dependent on the bend radius. The sensor shown at Fig. 9.10 takes advantage of this fact.

Any alteration in the spacing of the corrugated blocks shown, resulting from contact displacement, strain or pressure, will cause a change in the intensity of output light as a result of the variation of bend radius in the series of bends that the light must negotiate. This type of sensing device can be made very sensitive, capable of measuring displacements down upto 10^{-3} μm, strain upto 10^{-7}, and pressures upto 10^{-6} kg/mm^2. However, its dependence on variations in the source characteristics appears to be its main disadvantage.

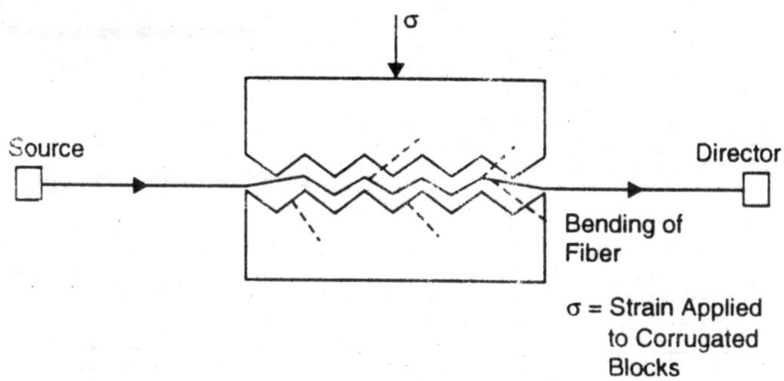

Fig. 9.10. Bend loss type strain sensor.

9.2.3. Fiber Optic Flow Transducer

In a sensing device in which bending of optical fiber is much less severe than the above, the light rays instead of getting lost, will get redistributed amongst the possible modes. This effect is utilized to measure the flow rate of a liquid in the configuration shown in Fig. 9.11. the optical fiber is stretched transversely. As the turbulance caused by it makes it to oscillate from side to side at a frequency roughly proportional to the flow rate. The result of this oscillation is a corresponding modulation of the mode power distribution. The intensity profile of the emerging light will be modulated accordingly. A photo detector that intercepts a fraction of the total emerging power will therefore provide an output which will

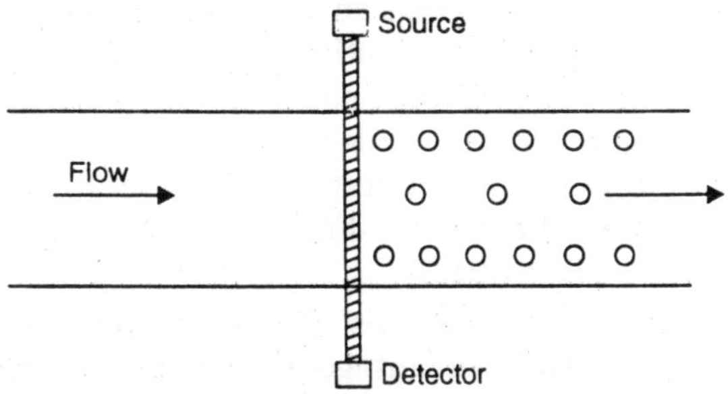

Fig. 9.11. Fiber optic flow rate sensor.

contain a component whose frequency is indicative of the flow rate. Such a device can measure flow rates from 0.3 m/s to 3 m/s. However such flow rates which are below the rate at which turbulance sets in, cannot be measured. Also, the flow rates that cause the optical fiber to vibrate at frequencies close to its natural resonance frequency, are not possible to be sensed.

9.2.4. Fiber Optic Sensors Based on Interferometry

The foregoing were rather simpler examples wherein the measurand was allowed to modulate the light level. The more sensitive and accurate methods involve the optical phase of light. Such methods are based either on interferometry or polarimetry.

The basic principle of interferometric optical fiber sensor is illustrated in Fig. 9.12. There are two single mode fiber paths, fed by a laser source. One of these is exposed to the measurand, its phase of light is altered and the output level resulting from the optical interference between the signals gets varied. This type of configuration is known as Mach-Zehnder type which is primarily a transmission-based interferometer involving two closely spaced optical fibers. The two arrows are to be matched in optical path to within the coherence length (or the distance over which the laser radiation remains in accurate step) of the source, so as to allow for direct use as an optical sensor.

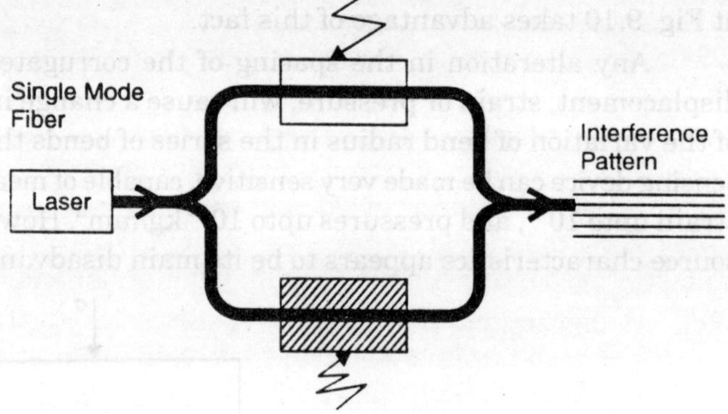

Fig. 9.12. Interference mode optical fiber sensor.

A very special form of Mach-Zehnder interferometer is the Fiber optic gyroscope and is one of the most highly developed one in this category. This is schematically shown in Fig. 9.13. The two phase arms are now within the same loop of the fiber, but are distinguishable by virtue of the fact that these propagate in opposite directions around the loop on emergence from their respective ends, the two signals are allowed to interfere. If the loop rotates about an axis through its centre and normal to its plane, one of the signals advances in phase (against the rotation) whilst the other (with the rotation) is retarded. The resultant perturbation of the interference pattern allows determination of the rotational velocity. This effect is known as SAGNAC effect (after the name of French scientist who first discovered it in 1913).

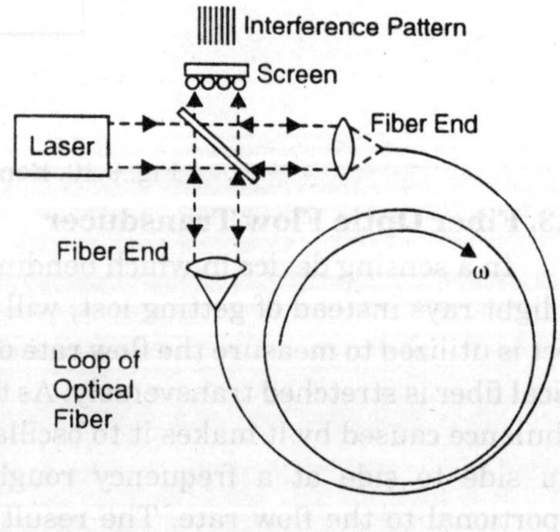

Fig. 9.13. Optical fiber gyroscope.

In order to establish the technique as a viable operational measurement method, it has to compete with advantage, the existing conventional techniques, such as "spinning inertial mass" gyroscopes. Consequently the demands made on the optical fiber version of groscope have become quite taxing ; for example, a sensitivity of rotational speed measurement of the order of

0.01° per hour. This requires reciprocity, *i.e.* ensuring that the two counter propagating paths are the same in every respect save that of the effect of rotation. This means that the two signals must have the same polarization, follow identical paths even outside the fiber, have the same power levels throughout and suffer no differential backscattering. The ultimate performance depends upon the extent to which these requirements are met. However, owing to the complexity of the engineering involved, the cost advantage is small. The advantages are more clearly evident for the less demanding applications, *e.g.* 0.1-10° per hour.

9.2.5. Distributed Optical Fiber Sensing (DOFS)

The optical fiber has a special property of making simultaneous measurements of a parameter field in both space and time domain. This leads to a new level of understanding as well as finer monitoring and control. DOFS takes advantage of two additional properties of the fiber: its one dimensional nature and its mechanical flexibility. With the aid of these properties, it becomes possible to detemine the value of a wanted parameter as a function of position, along the length of a suitably configured fiber ; with arbitrarily large spatial resolution, the normal temporal variation is determined simultaneously, from the time dependence of the signal.

Such a facility offers many attractive possibilities for industrial and research applications. The value of having access to spatial/temporal behaviour of strain and temperature in large, critical structures e.g. dams, air space craft, bridges, multi storey buildings, electricity generators, chemical pressure vessels etc., are clear from the points of view of safety monitoring and improved understanding of behaviour under anomalous conditions.

A distributed temperature sensor uses optical time domain reflectometry and Raman scattering to determine the temperature at any location along the fiber length, thus justifying the word "Distributed". Typically, a 1 km length of optical fiber can be used to measure temperature to ± 1°C resolving the distance less than 7.5 m. The technique consists of launching of short pulses of laser light into communication-grade optical fiber cable; as the light propagates through the fiber, static and dynamic variations cause scattering in all directions including back towards the source. Timing the arrival of scattered light gives the location, along the fiber, of the point being measured. Temperature measurement is possible with this sensor form – 173°C to + 400°C with a resolution of 0.5°C, while the spatial resolution could be as small as one metre.

However, these sensors are still at various stages of development and have good promise for future.

9.2.5.1. Fiber Optic Temperature Sensor

Some types of fiber optic sensors are still being used for temperature measurements. Temperatures over a wide range, namely from temperatures of cryogenic liquids to those of burning gases can be measured with good accuracy. One such sensor is shown in Fig. 9.14 which consists of a probe, a directional coupler, a laser which acts as a source of light and photo detector.

The probe consists of a polished tip of a fiber and a small mirror mounted on it. The fiber carries light radiation from laser to the tip. The inside surface of the fiber end reflects part of the light back into the fiber. The mirror reflects the remaining light transmitted through the tip also back into the fiber. The two reflected light beams interfere. The intensity of the light returning through the fiber varies with the phase difference between the reflected beams, which, in turn, varies with the distance between the fiber and the mirror surface. This gap changes with temperature as a result of thermal expansion and concentration of the probe. A directional coupler extracts part of the returned light from the fiber and channels it into a second fiber. A photodetector converts the extracted light into an electrical signal which is a measure of the probe temperature.

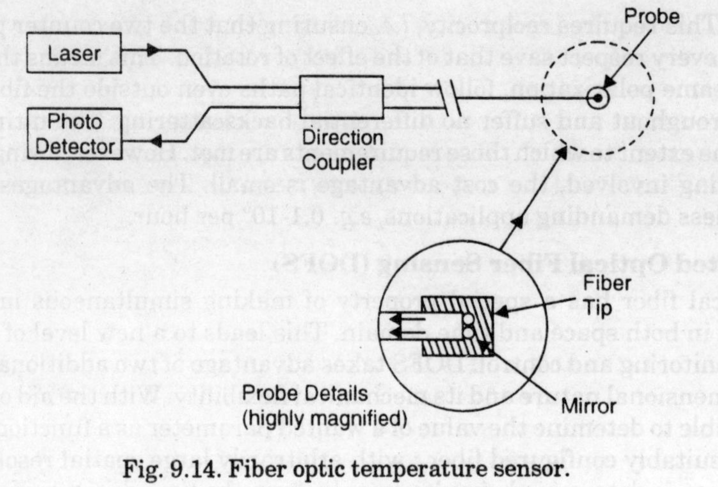

Fig. 9.14. Fiber optic temperature sensor.

This type of sensor is light weight, even lighter than a thermocouple, it is immune to electromagnetic interference. Also, the sensor would not respond to the temperatures other than those at the sensing tip.

9.2.5.2. Optical Time Domain Reflectometry (OTDR)

In connection with the distributed Optical fiber sensing (DOFS), a mention was made of the optical time domain reflectometry. This basic technique requires some elaboration. In the arrangement shown in the Fig. 9.15, a short pulse of light is launched into a single mode optical fiber and as it propagates along the fiber length, a very small fraction of light will be continuously scattered back towards the launching end by small imperfections in the fiber material. Since the velocity of light in the fiber is known (about 2×10^8m/sec), it is possible, by noting how the returning level of light varies with time, to determine from which point in the fiber each portion of light has been returned. This technique entirely analogous to Radar, is known as OTDR and is used in telecommunications to locate fiber breaks, bad joints or very loss producing fiber sections, if, however, we also note how polarization state of the returning light varies with time. This tells us how the polarization properties of fiber vary along its length, and if some external electric or magnetic field is acting on the fiber to alter its polarization properties then we have a method of measuring the distribution of the external field along the fiber. This technique, known as Polarization Optical time domain reflectometry could be a very powerful probe for determining the way in which various important quantities varied with position, as well as with time.

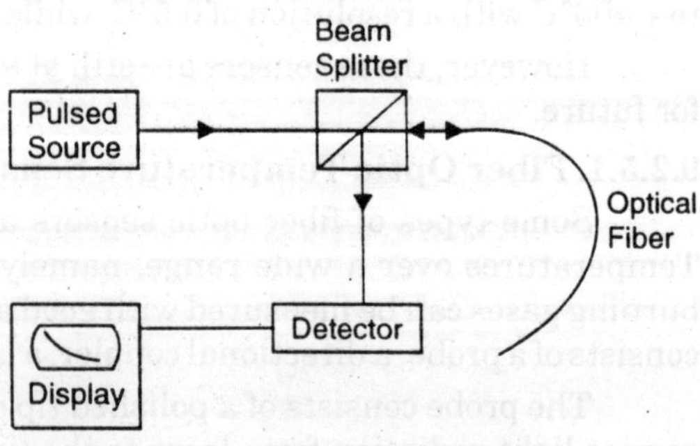

Fig. 9.15. Optical time domain reflectometry (OTDR).

9.2.5.3. A Simple Fiber Optic Pressure Sensor

As mentioned above, optical fiber sensors have been used in intrinsic mode where modulation of optical phase, wavelength, intensity or time may be caused by the measurand without interrupting the fiber path. Alternatively, they may also be used in extrinsic mode where the measureand interacts with the light externally from the fiber path. The sensors based on intensity modulation are the most attractive mainly because of their simplicity, high resolution and sensitivity.

Pressure being one of the mechanical parameters is measured by transduction in various ways. Most pressure transducers use diaphragm as an elastic member at the primary stage of pressure sensing because of its simplicity and freedom from hysteresis. The most suitable material for diaphragm is stainless steel or molybdenum.

The pressure cell is made out of a solid brass cylinder about 60 mm outer diameter and 45 mm in total depth, closed at one end by a welded stainless steel flat diaphragm. A very thin mirror like aluminised circular sheet of plastic with a diameter 1 cm is fixed at the diaphragm centre so as to serve as a mirror for reflecting light from the transmitting fiber. The length of fibers from the end to the 'Y' joint is 15 cm each and the end terminals of each fiber has an aperture 3 mm in diameter. A 3 V torch lamp can be used as a source of light fixed at one end with the help of a connector and a covered light tight selenium photovoltaic cell as a photo detector, connected to the return fiber end. The 'Y' joint is fixed through a connector on a screwable cap covering the diaphragm as shown in Fig. 9.16 (*a*). The electrical output can be measured by a digital voltmeter. When pressure is applied through the tapered inlet, the diaphragm gets displaced causing modulation of

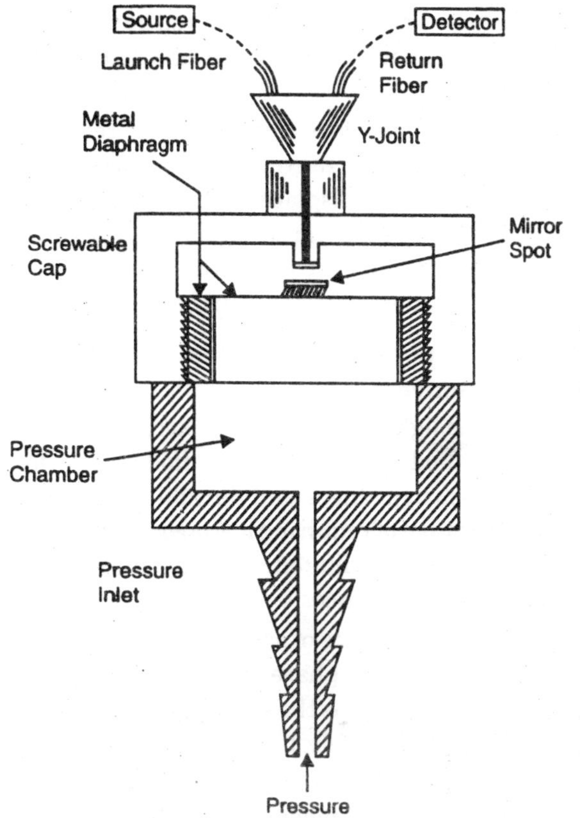

Fig. 9.16. (*a*) Pressure cell assembly.

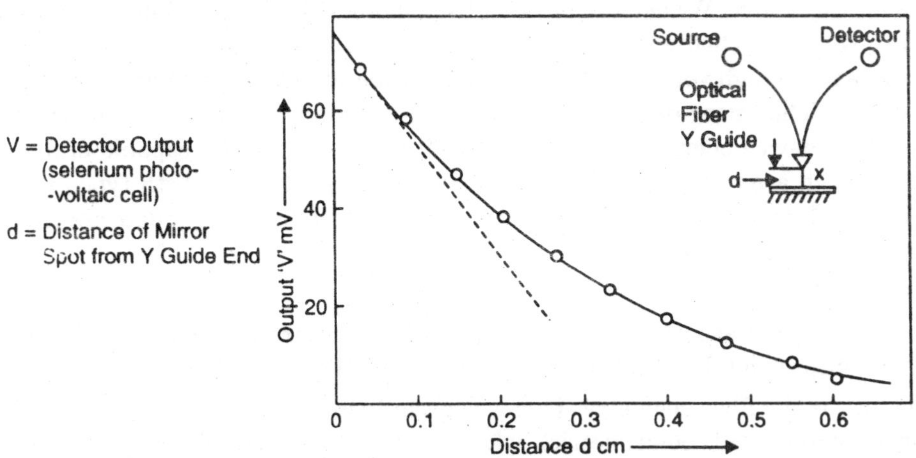

Fig. 9.16. (*b*) Calibration curve.

light intensity reaching the photo detector. The output signal after amplification, can be sent to the photovoltaic cell.

For calibration, the 'Y' guide is mounted on a fine screw vertical movement with a 10 mm least count. The output is measured under dark room conditions by varying the disatnce between the mirror spot on the diaphragm and the 'Y' joint end. The calibration curve is shown in Fig. 9.16 (*b*).

The linear variations occur for relative displacement of about 500 mm. However, for a flat diaphragm, the maximum tolerable deflection is about 30% of its thickness for maintaining a linear pressure-displacement relationship. The calibration curve also indicates that output voltage Vs distance variation does not obey inverse-square law. Therefore, such a transducer needs calibration over the entire range of displacement, with the given constant output light source and photo detector. Further improvements and modifications in the device are required.

9.2.6. Components for Optical Fiber Sensors

9.2.6.1. Sources

The light source in any practical industrial system has to be either a semiconductor light emitting diode (LED) or a laser diode (LD). Only a packaged semiconductor module is expected to possess ruggedness and durability required in hostile environments. In addition, it enjoys the advantage of requiring low operating voltage. The most severe demands on sources are those made by the distributed systems, since these have special requirements in respect of pulse energy and coherence.

The coherence of source is required for any DOFS method which utilizes phase modulation. Therefore, for phase information to be extracted satisfactorily, the coherence length must be at least as large as the spatial resolution. Thus, a coherence length of > 0.2 m and a coherence time of approximately 1 ns are required.

This translates into a source spectral linewidth of < 1 GHz and can be achieved only with single longitudinal mode semi-conductor lasers.

9.2.6.2. Detectors

For detection of optical radiation, we require a semi-conductor device, for the same reasons as for the source. We can use either a PIN photodiode or an avalanche photodiode. For back scatter, we would choose the latter for its increased sensitivity. Its drawbacks are that it requires a relatively high voltage (100-200 V) for its operation and is intrinsically noisier than the PIN device. The latter device would be quite adequate for forward scatter distributed systems as PIN device operates at 10-20 V. Both types are readily capable of about 1 GHz bandwidths which are required for DOFS function. A large range of both types are available commercially, in both silicon and germanium structures, and the choice of detector thus presents relatively very few difficulties.

9.2.6.3. Optical Fibers

There are three aspects of fiber composition which are important in the provision of optical fiber sensors and these are common to both DOFS and point systems :

The first is the fiber geometry. This is especially important in relation to phase devices (*i.e.* interferometric and polarimetric) where controlled polarization characteristics are desirable. One geometrical approach is that where dual cores are positioned within the same cladding. Such a structure allows the measurand to control the coupling between the two guides and thus effectively separates the transmission and receiver channels. This has distinct advantages in detection.

The second important aspect of fiber fabrication is that of dopant compostion. The inclusion of dopants can render the fiber's optical properties specially sensitive to a particular measurand and thereby ease very considerably the DOFS design. Some progress has been made in this area, particularly in regard to temperature measurement. A number of rare earth elements have been tried with interesting resulting values for temperature coefficients of absorption, especially in case of Holmium. Other properties of these fibers are under intensive investigation. These include non linear properties such as those that might lead to enhancement of electro optic effects, Raman scattering coefficients etc. These latter developments are unlikely to have any direct impact on the measurement fiber since even with enhancement, the non linear effects are still likely to require much higher peak optical power levels than can readily be provided by semi conductor lasers. They might very well give rise to a range of active fiber components (*e.g.* phase/frequency modulators, polarizers, amplifiers etc.), which could prove invaluable in the signal processing functions necessary for optimal extraction of the measurand information in DOFS.

There is very much more that could be done in the area of dopants. Dopants which could lead to some controlled anisotropy in the fiber, for example, so as to enhance elasto optic effects for strain/pressure measurements and so as to allow enhancement of measurand-dependent mode coupling in the fiber.

Thirdly, the fiber coatings. Very little attention has yet been devoted to the coating problems which are thrown up by sensor requirements.

Firstly, it is necessary to ensure that the coating can survive any mechanical or chemical attack to which it might be subjected in the measurement environment; secondly, the coating must allow optimum interaction of the measurand with the fiber property which is being modulated in the measurement action; and thridly, that interaction must remain constant over long periods of time if frequent, inconvenient (and possibly unallowable) recalibration is to be avoided. The coating problem is not therefore, at all an easy one. A range of measurements and applications, in fact, pose a series of problems with regard to coating and this alongwith with dopant area, call for a concerted attack by material scientists and engineers.

It must be emphasised that the subject of optical fiber sensors has progressed to its present state largely by using fibers designed for telecommunication purposes, (since the commercial prospects in this latter area are perceived to be much more profitable). Perhaps, it will become widely accepted for industrial application only when the special problems appropriate to sensors and transducers application are properly addressed and solved.

BIBLIOGRAPHY

1. W. Gopal, J. Hesse & J.N. Zemel Editors, *Sensors: A Comprehensive Survey: vol. 6 : Optical Sensors*, Weinheim (Germany, USA), 1992.
2. V.R. Moorthi, Pramod Kumar and Arvind Gupta, *Transducers and Sensors for Distributed Control Systems*, IETE Technical Review, 9(2) March-April 1992, p. 177.
3. S.K. Chakravarti, *On the Design of a Simple Pressure Cell Using Optical Fiber Sensor*, IETE Technical Review, 9(2) March April 1992, p. 184.

4. Brian Culshaw and John Dakin Editors, *Optical Fiber Sensors: Systems and Applications*, Artech House Inc. 1989.
5. H.J. Arditty, J.P. Dakin and R. Th. Kersten, Editors, *Optical Fiber Sensors*, Proc. of 6th International Conf. Paris, 18-20 Sept. 1989, Springer and Verlag 1989.
6. T.R. Padmanabhan, *Industrial Instrumentation: Principles and Design*, Springer-Verlag 2000.
7. D.V.S. Murthy, *Transducers for Instrumentation Systems*, Prentice Hall of India Ltd. 1995.
8. S. Soloman, *Sensors and Control Systems in Manufacturing*, Mcgraw Hill International 1994.
9. D.P Dewitt and G.D. Nutter, *Theory and Practice of Radiation Thermometry*, Wiley, New York 1998.
10. D. Patranabis, *Sensors and Transducers*, Wheeler, 1997.

10

Digital Transducers

The ease and versatility provided by digital signal processing circuits and the digital computing necessitates transducers providing digital output. One approach is having signals in the analog form and using A/D convertors to obtain digital outputs. Another approach is using direct digital transducers which provide output signals in the form of rectangular pulses of constant duration and amplitude, the presence or absence of a pulse in its time slot is taken to stand for either 1s or 0s in binary notation.

Encoders are most commonly used digital transducers. These are used in a variety of applications, *e.g.* machine tools, robotics and gauging etc.

10.1. ENCODERS

All encoders require a system of sensing which could be either contacting type, using mechanical brushes, magnetised strips or conducting strips. However, all encoders are invariably used for measuring displacement either linear or rotary (angular). A typical linear digital encoder (LDE) is illustrated in Fig. 10.1.

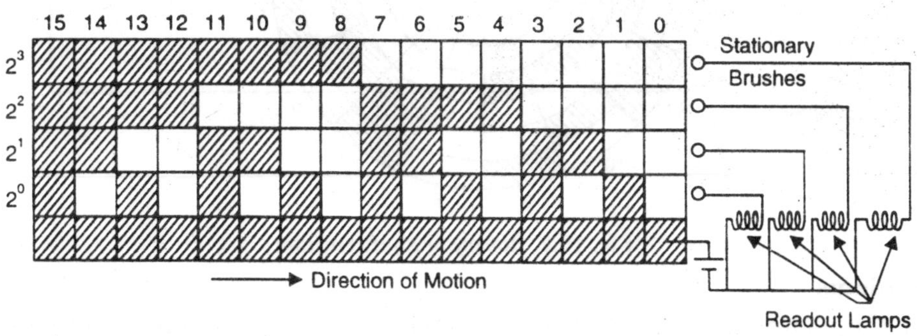

Fig. 10.1. Linear digital encoder.

10.1.1. Linear Encoders

As shown in the Fig. 10.1, the length of encoder is 16 times the length of traverse required for a change of digit in the least significant bit (LSB). If a resolution of 0.2 mm is required, a maximum traverse of encoder in the direction of motion is $15 \times 0.2 = 3$ mm. Brushes make contact over an area of each track much less than 0.2×0.2 mm. The scale of encoder moves under four tracks. The respective lamp circuits are connected or broken so that the position of scale is readout by the lamps under glow. Thus 0000 will give no glow and 1111 will make all lamps glow.

If a traverse of longer than 3 mm is required, then the number of tracks must be increased, in which case, much finer brushes are required. Limitation is generally set by the sensing or scanning system.

Disadvantages of brush pick ups are :

- brush contact
- brush bounce and
- brush friction.

10.1.2. Shaft Encoders

Shaft encoders are used for measuring angular or rotary displacement. These are available in the form of discs having multiple tracks. The principle remains basically the same as in case of linear encoder. Discs upto 10 tracks are available. A 10 track disc has a resolution of 1 in 1024(2^{10}).

This is illustrated in Fig. 10.2.

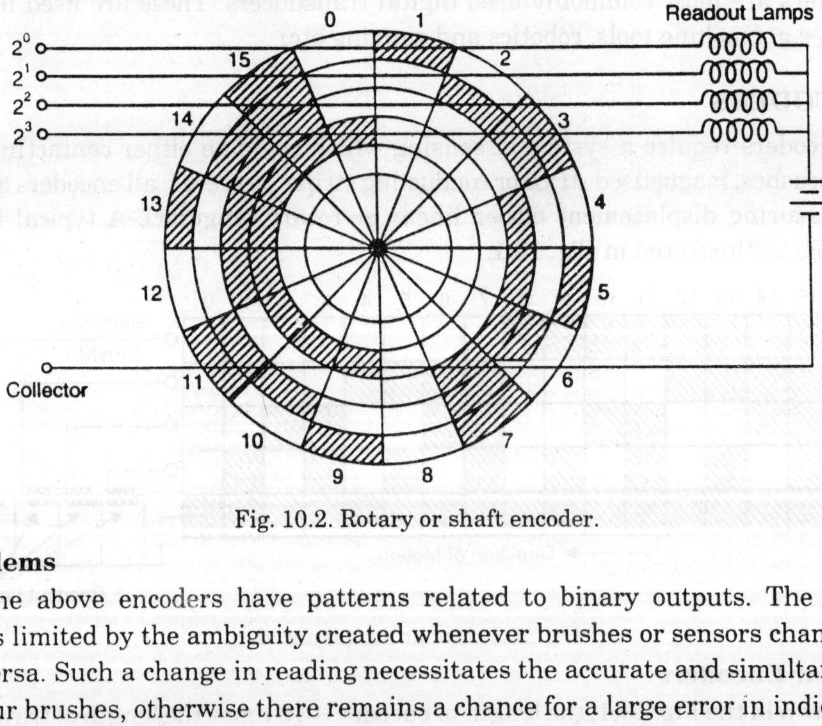

Fig. 10.2. Rotary or shaft encoder.

Scan Problems

All the above encoders have patterns related to binary outputs. The reliability of indication is limited by the ambiguity created whenever brushes or sensors change from '1' to '0' or vice versa. Such a change in reading necessitates the accurate and simultaneous change of all the four brushes, otherwise there remains a chance for a large error in indication.

In a progressive or reflected code, only one digit is changed at a time. For any transition point between 0 to 15, the readout encounters a change in one brush position only. The code requires conversion to pure binary and involves utilization of a Gray-to-Binary decoder based on exclusive OR gates as shown in Fig. 10.3. Most significant bit (MSB) remains unchanged during conversion or decoding.

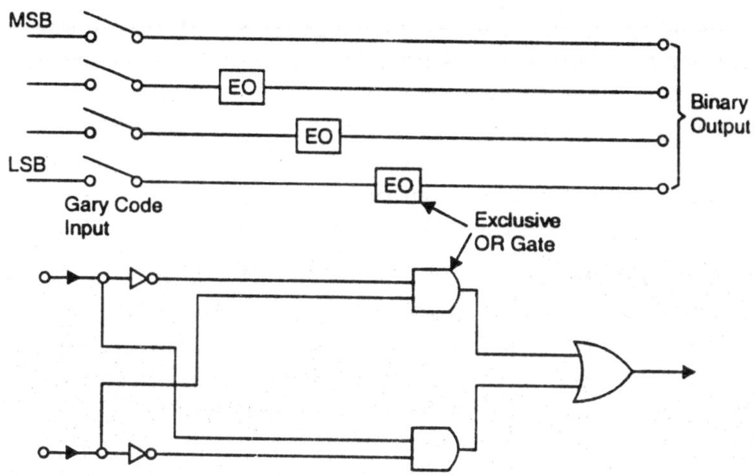

Fig. 10.3. Gray-to-Binary decoder based on exclusive OR gate.

A table giving both equivalent gray code and binary code (4-bit case) is given below :

Decimal Number	Binary Code	Equivalent Gray Code
00	0000	0000
01	0001	0001
02	0010	0011
03	0011	0010
04	0100	0110
05	0101	0111
06	0110	0101
07	0111	0100
08	1000	1100
09	1001	1101
10	1010	1111
11	1011	1110
12	1100	1010
13	1101	1011
14	1110	1001
15	1111	1000

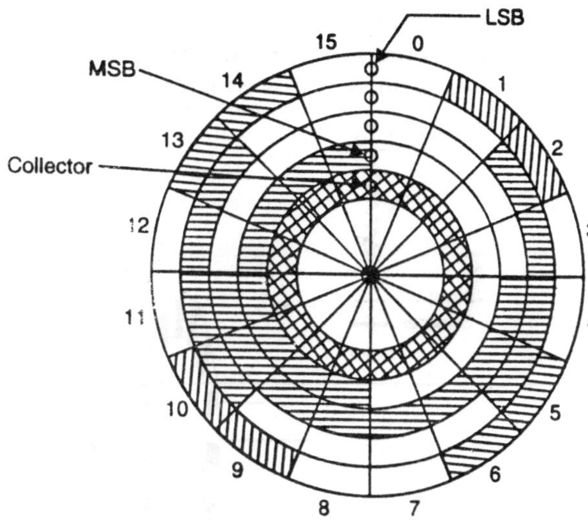

Fig. 10.4. Encoder based on 4-digit gray code.

An encoder based on a 4 bit gray code is shown in Fig. 10.4.

10.2. OPTICAL ENCODERS

These encoders employ a pulse disc firmly attached to a shaft. Based on their function, optical encoder can be either incremental or absolute.

With associated photoelectric detectors, one pulse in case of incremental or one code in case of absolute encoder is obtained for each position.

These encoders have a long life but their resolution is limited. The main advantages of optical encoders are high accuracy and virtual absence of mechanical wear in operation due to their non contact nature. A 12 bit optical encoder will have an inherent reading accuracy of $(1/2)^{12}$ or 1/4096 = 0.024%. Optical encoders are widely used as position transducers in robot systems.

However, the technique requires prior knowledge of the total number of lines on the disk so as to make the counting hardware keep track of the number of complete shaft revolutions. The changing of position measurement resolution which is determined by changing the total number of lines on disk, would also require a change in the counting hardware.

An alternative adopted by most major manufacturers, is the use of a third LED-photodiode pair and a separate channel containing a single opaque line (or clear slot in case of a metal disk), which then provides the detection of each complete revolution, as required.

10.2.1. Incremental Encoders

Both types of optical encoders transform mechanical input, namely the angular position of shaft into an electrical output by light absorption. Incremental encoders generate a certain number of pulses per revolution. They provide a pulse for each increment corresponding to the resolution but do not make distinction between two consecutive increments. Thus no identification for absolute increment is available. Incremental encoders are simple in construction, are low cost and easy to use. The number of pulses generated by an incremental encoder corresponds to a certain distance movement. A coded disc is mounted on the shaft and is divided into separate segments (gratings) which are alternately transparent or opaque. An LED emits a light beam that illuminates all the segments of the coded disc (Fig. 10.5). Photo-elements receive modulated light and convert the same into two sinusoidal signals which are phase shifted by 90°. Internal digitising circuit amplifies these signals (through an operational amplifier) and shapes them into square wave pulse trains which are transmitted via the cable driver at the output.

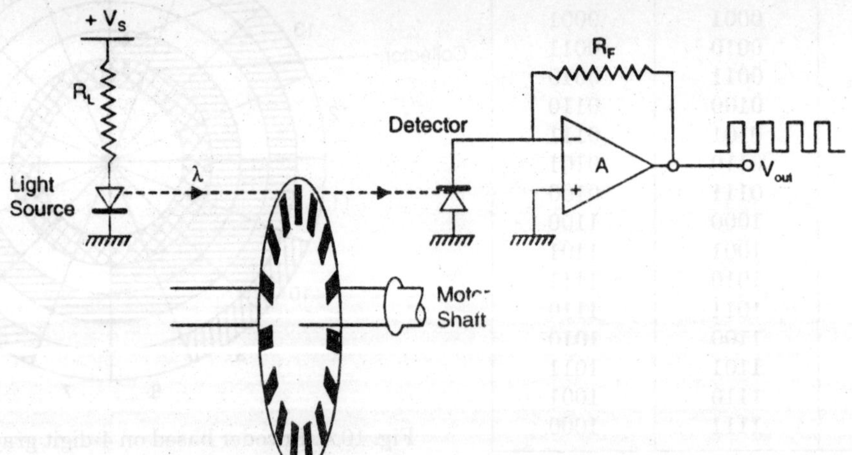

Fig. 10.5. Incremental optical encoder used as a position transducer.

Normally, incremental encoders give output signals (A, B, 0—index). These are generally available in 5V (TTL) design or 10-30 V design. Phase difference between signals A and B is 90°. These have a maximum switching frequency of 300 KHz.

Their main drawback is the poor angular resolution, $\Delta\alpha$ as shown in Fig. 10.6. This depends on the number of opaque lines on the disc denoted by n and the width Wp of the photodiode active area on the disk, which in turn depends on the photodiode mounting distance from the disc centre.

$$\Delta\alpha = 360/n$$

and the active area,

$$W_p = r \sin \Delta\alpha = r \sin 360/n$$

The square root of this will give the width W_p.

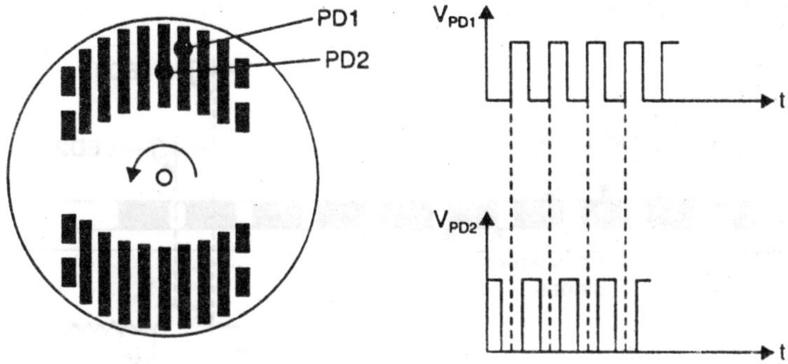

Fig. 10.6. (*a*) Positioning of angular direction sensing photodiodes on encoder disc.

For example, to design an optical incremental position transducer with a resolution of ± 3° would require an encoder disk with at least 120 slots, one in every 3° of arc. To match this resolution, the photodiode would need to have an active width W_p not bigger than 1/4 mm^2 mounted 10 mm away from the disk centre.

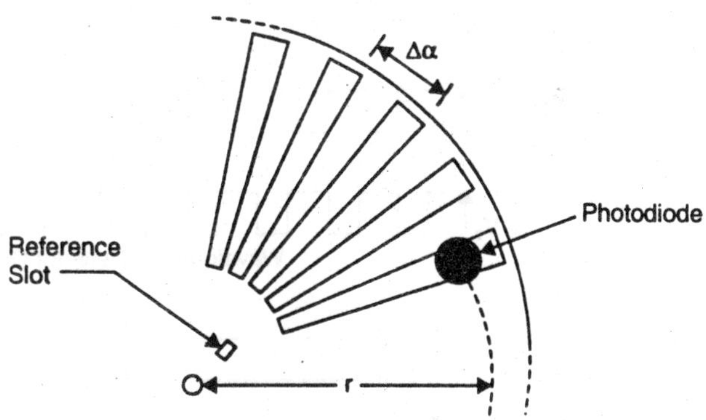

Fig. 10.6. (*b*) Angular resolution of incremental encoder.

One way to increase optical resolution is to use a second stationary plate called phase plate between the rotating disk and the photodiode so as to produce Moire' interference frings on the photodiode surfaces. Such transducers are commercially available. The angular resolution of this does not depend on the photodiode active area but on the width of the Moire' fringes line width which can be adjusted to suit the particular application.

Even filament lamps could be used as a light source instead of LED. Though LEDs have a longer life and great immunity against mechanical disturbances, their efficiency of light output is far less than the filament lamps. Alongwith filament lamps, silicon photovoltaic cells developing an output voltage of 20-40 mv in a load resistor of 10 kΩ are used as sensors/detectors.

Incremental Encoder for Linear Displacement

For linear displacement, an incremental encoder uses a strip of alternate stripes with an LED and phototransistor. An increment in displacement in either direction gives one output pulse. However, to detect the direction of motion or displacement, another LED and phototransistor is used alongwith the strip. The second sensor is displaced a 1/4 cycle from the first one. It yields another 0-1 output sequence 1/4 cycle shifted in phase from the first one. In the arrangement shown in Fig. 10.7, the pairs LED_1, T_1 and LED_2, T_2 can be displaced in phase by 1/4, 5/4, 9/4 cycles etc. A normal case of bright stripe : dark stripe ratio of 1/3 : 2/3 has been shown. Depending upon the direction of displacement of strip, outputs at points 1 and 2 are as shown. Output V_1 lags that of V_2 by a quarter cycle. These go as the D input and the clock input respectively to the D flip flop. The output from the flip flop is 1 for the direction of motion or displacement shown and zero for the opposite direction. Thus Q gives the direction of motion and 1 or 2, the quantum change.

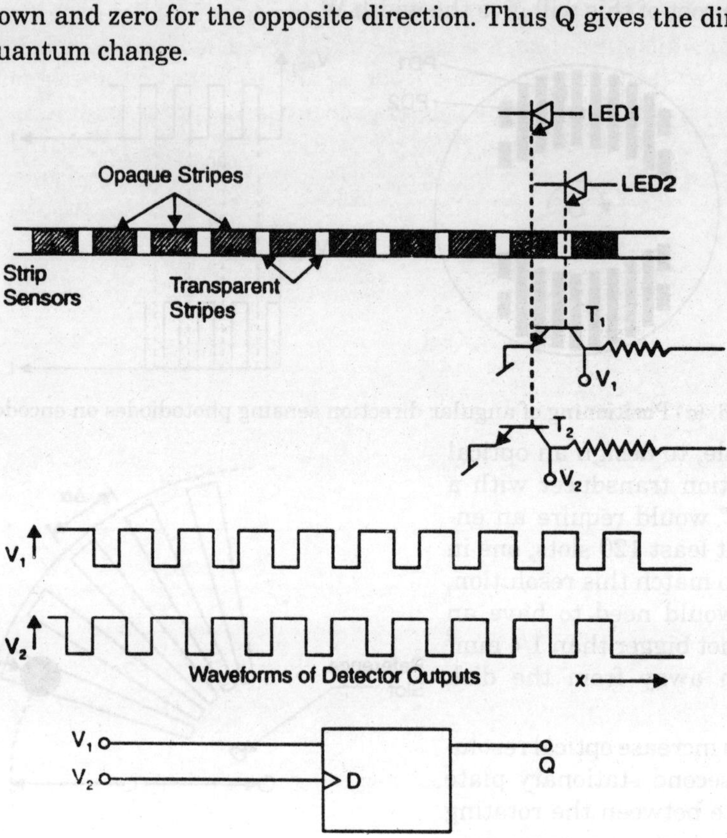

Fig. 10.7. Incremental optical encoder schemes.

The pulse train available from the encoder is counted using up and down counter. Counting up or counting down is decided by the Q output from the D flip flop. It is thus possible to get a cumulative value of displacement at any time. To fix an absolute reference point or zero of the scale, the strip is provided with a separate dark spot with a dedicated LED-phototransistor pair for it. When this pair has dark spots across, the position is taken as zero. The up-down counter is cleared and counting started afresh. With these two changes, sensing of absolute position and direction of motion are effected.

10.2.2. Absolute Encoders

In incremental encoders, a separate zero position detector decides the value for displacent. Whenever, power supply to it is made on or interrupted, it is to be reset. This frequent resetting is required to be avoided in high flexibility and high reliability position sensing systems. Absolute encoder provides the digital output corresponding to the position directly. Each bit position is separately encoded through a dedicated LED- photodetector pair to detect the logic state of each bit.

Absolute encoders therefore provide a distinct digital code indicative of each particular significant increment. However, these encoders become complex, costly and require special evaluation electronics for encoding position.

The coded value at the output is available immediately once power is applied. A coded disk is firmly mounted on the shaft. This has several tracks/segments of line markings. When being scanned in parallel, they provide encoded angular data. The light source emits a parallel oriented light beam which illuminates all segments of the coded disk. Photoelectric conversion process is identical with the incremental encoders. Fig. 10.8 shows the schematic of internals of an absolute encoder. Their switching frequencies are of the order of 50 KHz.

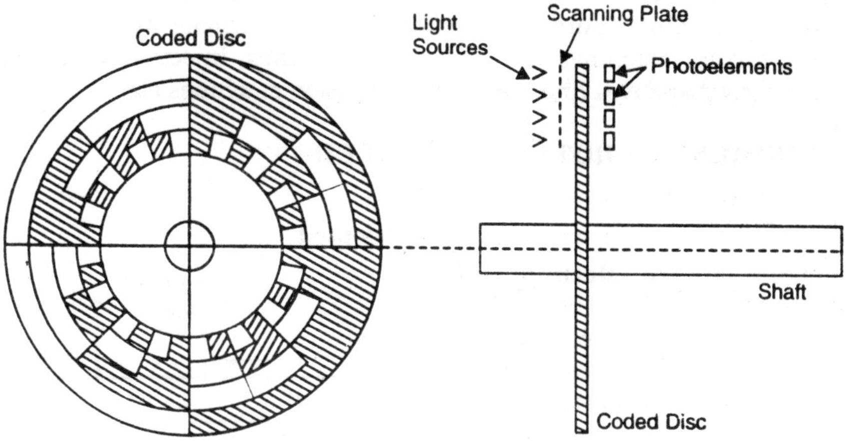

Fig. 10.8. Schematic of the internals of absolute encoder.

Conventional absolute encoders are of single turn with one revolution of 360° encoded in n increments and then the codes repeat beyond one revolution. The output is provided directly as a digital number representing the angular position measurement whose resolution depends on the number of bits, that is the number of channels used, as shown by the following equation:

Required angular resolution = $360/2^c$

c denotes number of channels required.

For example, to design absolute encoder with a resolution of less than 5° of arc, one would need an encoder disc with 7 channels, since $360/2^7 = 2.81°$, which is physically bigger and more expensive than an incremental encoder of comparable resolution.

The encoded disk pattern usually employed is based on a gray code because it minimizes measurement errors since it allows only one bit to change state at any point in time. The alternative, namely a binary coded disk pattern, is not very popular because it may produce 'race' conditions in the digital counting hardware which could lead to transition errors in measurement. This problem can occur in digital systems whenever more than one logic state changes at the same time.

However, the absolute encoders and position transducers have been restricted in their use to high reliability, non-space-critical applications in view of their larger size and cost.

Service life of encoder is determined mainly by life of the bearings and light sources used. Generally, ball bearings and Gallium Arsenide light source is preferred.

Optical encoders are very sensitive devices. Rugged mounting is extremely necessary to exploit desired resolution and accuracies offered by these. They must be protected against mechanical overload, including displacement between the drive and the encoder shafts, to avoid damage to the shafts and bearings and errors in measurement.

Gray code output is popular in encoders. However, some encoders offer BCD or binary code output also. Gray code output is preferred when data is to be transmitted over long distances. This is because there is only single bit code change between adjacent resolution positions eliminating abrupt change in output data reception. Binary or BCD code output is preferred when other control electronics is present nearby encoder only and the environment is less polluted by noise.

Availability of reliable encoders has resulted in their usage almost in every sphere of automation. Encoders are used in applications like door monitoring machines, table robots, plotting and testing machines, automated screw driver, ultrasonic welding machines, CNC machines, lathes, mixing machines, wire drawing machines, injection moulding machines, welding robots, spinning and weaving machines, extrusion machines etc.

10.3. MEASUREMENT OF ROTATIONAL SPEED

10.3.1. Pulsed Output Sensors

The following methods are most commonly adopted :

A. Electromagnetic Method. As shown in Fig. 10.9, when a toothed iron wheel rotates, it varies the reluctance between the wheel and the ferrite core, periodically. Thus a pulsed voltage signal is generated in the pick up coil which is proportional to the number of teeth in the wheel. This is obviously, a measure of rotational speed. However, the device is unsuitable for detecting low speed because the output voltage pulse is too small. However, the excellent durability and not requiring a separate power source are the advantages.

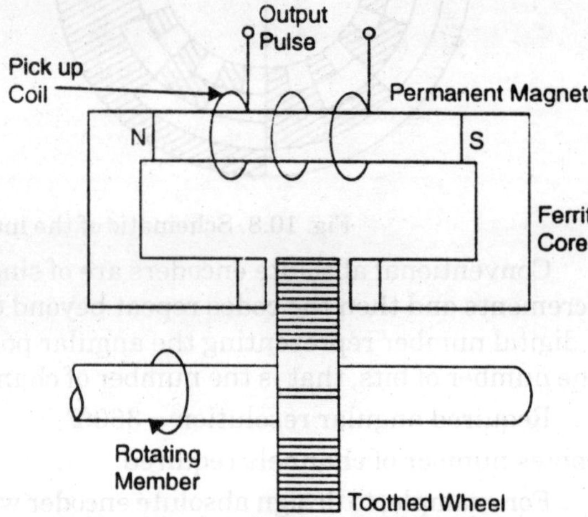

Fig. 10.9. Electromagnetic pulsed output sensor.

B. Photoelectric Method. As a slotted disc rotates, it interrupts the incident light from a filament lamp source and the photocell detector produces an output voltage pulse which is proportional to the rotational speed. Resolution can be increased by increasing the number of slots on the rotating disc.

The schematic arrangement is shown in Fig. 10.10.

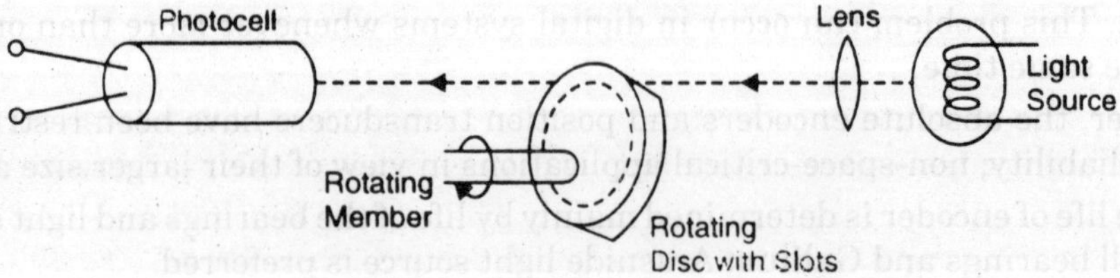

Fig. 10.10. Incident light pulsed output sensor.

Another arrangement makes use of reflected light in place of direct incident light, as shown in Fig. 10.11. The rotating disc has black and white areas around the periphery. The photo cell or photo transistor receives output in the form of pulsed signals generated by the light reflected back to it from the black and white areas on the disc in an on/off fashion.

Since there is no contact with the rotating member, it is used for detecting high rotational speeds. The method places no load on the rotor. It is therefore suitable for measuring low torque rotation speeds.

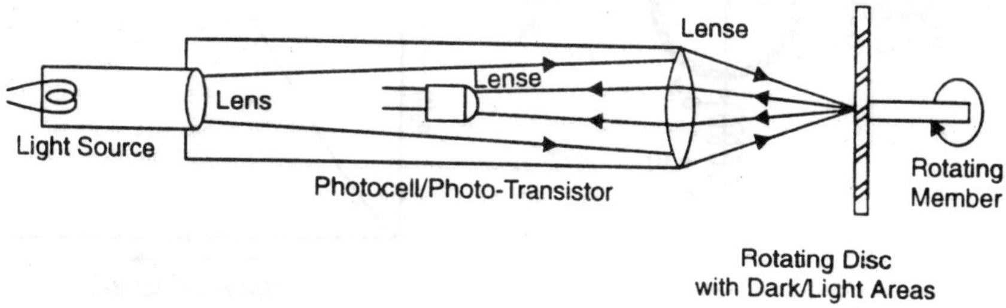

Fig. 10.11. Reflected light pulsed output sensor.

10.3.2. Digital Counting Techniques

One method is illustrated in Fig. 10.12. It directly counts the number of pulsed signals in a fixed amount of time. The time duration T is determined by a stable quartz oscillator. The gate opens for a fixed period of time by a control signal whose duration is measured.

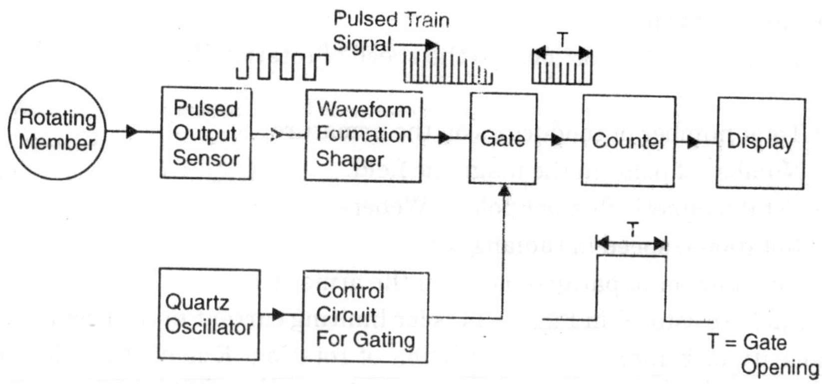

Fig. 10.12. Block diagram of digital pulse counting technique.

10.4. TACHNOGENERATORS

10.4.1. AC Tachogenerator

AC tachogenerator is essentially a modified version of an induction motor. Both rotor surface and the inner stator surface are cylindrical. The air gap is of the order of 0.2 mm. The stator carries a two-phase winding while the rotor is a "Squirrel Cage" one, arranged in slots around the periphery. The stator windings are uniformly distributed and are 90° apart. One winding is excited at the carrier frequency and synchronous speed. The rotation causes dynamically induced emfs in the rotor which are in phase with the flux Φ_d and hence proportional to speed of rotation. Resulting currents in the shorted rotor conductors produce an alternating magnetic flux Φ_c at 90° to Φ_d. It is this cross-flux Φ_c that induces voltage in the output winding.

The AC tachogenerator shown in Fig. 10.13 (*a*), provides a high sensitivity, 0.1 V/rpm or even more.

Carrier frequencies of 50 Hz, 60 Hz and 400 Hz are common. However, sensitivity gets affected by the changes in amplitude of excitation. The variation of output voltage $e_0(t)$ with speed of rotation V is shown in Fig. 10.13 (*b*).

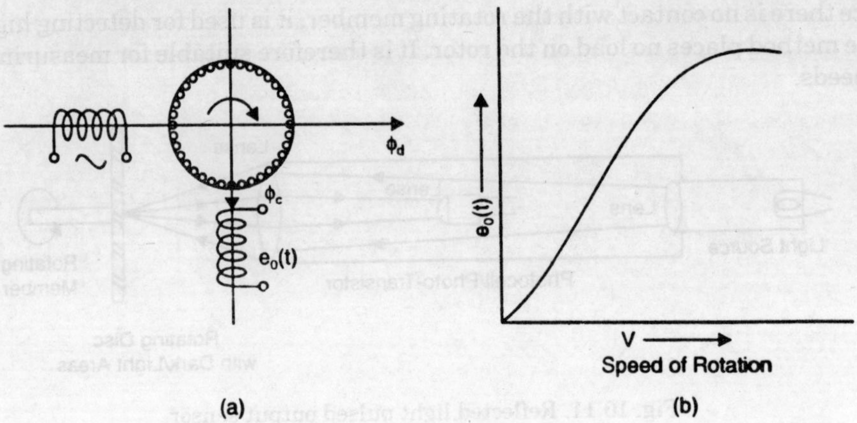

Fig. 10.13. A.C. tachogenerator.

10.4.2. DC Tachogenerator

A DC tachogenerator is a miniature DC generator with a permanent magnet stator and a wound armature. The impedance or resistance of the winding is kept low for good sensitivity. The emf generated is given by

$E = Np\Phi V/a$, and is proportional to the speed of rotation V.

Here :

N = Total number of conductors on the armature

P = Number of poles in the magnetic field.

Φ = Total magnetic flux per pole in Webers

V = Rotational speed in radians/sec.

a = The number of parallel paths on the armature.

The variable resistor R in Fig. 10.14 is for limiting current in the event of a short circuit.

The polarity of E indicates the direction of rotation. E is of the order of 5-20 mv per revolution. The periodical maintenance of commutator and brushes is required and fine brushes are preferred.

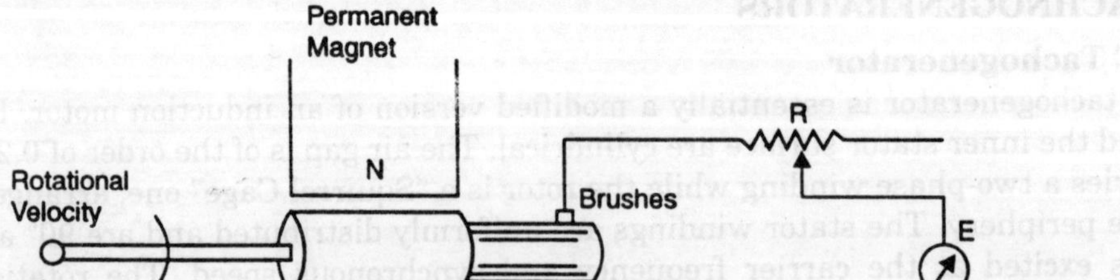

Fig. 10.14. D.C. tachogenerator.

One advantage of DC tachogenerator is that the speed vs output voltage relationship remains linear over a wide range. The ripple contact in the output changes with rotational speed. It may be quite high at low speeds and this sometimes poses problems.

However, a tachogenerator is essentially an analog device and the output from it requires A/D conversion and in some applications, low pass filter to commutator spikes (in case of DC tachogenerator) and/or output voltage attenuation to match it to the A/D working range. An arrangement for interfacing is shown in Fig. 10.15.

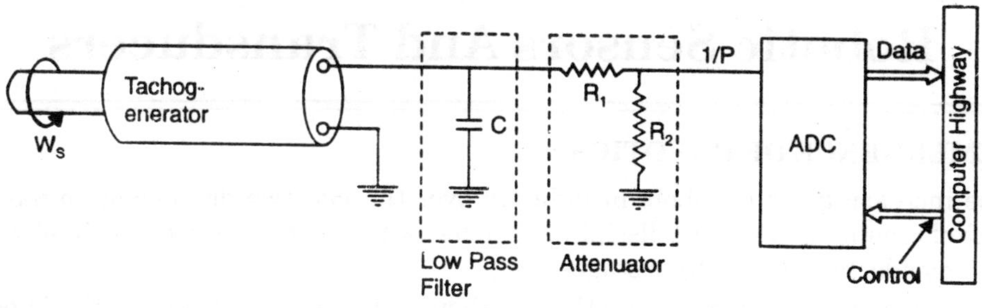

Fig. 10.15. A generalised computer interface circuit for a tachogenerator.

Because of the analog output, the speed or velocity measurement resolution depends only on the number of bits used by the A/D converter ; an 8-bit converter will therefore provide just under 0.4% measurement resolution, that is $1/256 \times 100 = 0.39\%$.

BIBLIOGRAPHY

1. D.V.S. Murthy, *Transducers and Instrumentation Systems*, Prentice Hall of India Ltd. 1995.
2. C.S. Rangan, G.R. Sarma and V.S.V. Mani, *Instrumentation : Devices and Systems*, Tata Mcgraw Hill, 1983.
3. TR Padmanabhan, *Industrial Instrumentation : Principles and Design*, Springer-Verlag 2000.
4. S.R. Ruocco, *Robot Sensors and Transducers*, Halsted Press, John Wiley New York, 1987.
5. R.G. Seippel, *Transducers, Sensors and Detectors*, Reston Publishing Co. 1983.
6. Orlosky S. Avolio G., *Basics of Rotary Optical Encoders*, Control Engineering, Nov. 1997.
7. V.R. Shah and V.R. Kanetkar, *Rotary Position Sensors : An overview*, Proc. National Conf. on Sensors and Transducers held as CSIO, Chandigarh on 20-21 Sept. 1996, p. 71.

11

Robotic Sensors And Transducers

11.1. EMERGENCE OF ROBOTICS

The increasing use of robots in industry over the last two decades or so has seen the emergence of robotics as a subject discipline with its own identity. The robot technology has also evolved alongside, broadly in three stages:

The first generation robots were those without any sensors or transducers. These did not have computing power to interact with their environment. The control of end arm was effected by calculating its position with the help of data supplied by position transducers placed within each robot joint. Such position transducers have been described in details under the head "Encoders" in chapter ten.

The second generation robots have some limited computing power which allows them to process the environmental data in a limited way.

While the first generation robots are still being used commercially, the second generation ones are still largely at an experimental stage of development and use.

The third generation robots are equipped with sensors and have extensive computing power to enable them to make decisions, plan and execute the tasks required. Such robots are yet to enter the market, in a big way.

The second and third generation robots therefore require sensors that can provide the necessary environmental feedback and help to make them more accurate and flexible. A long term strategy of developing a general purpose and flexible sensor is ultimately suited to the flexible manufacturing systems which require large investments. This type of research effort therefore tends to be limited to academic institutions and large corporate sector. The use of sensor with machines whether to control them continuously or to inspect and verify their operation, can be highly cost effective in particular areas of industrial automation. Examples of such areas include sensing systems to monitor tool condition, force and torque sensing for robot assembly systems, vision based automatic inspection, and tracking sensor for robot arc welding and seam sealing. Many of these will be the basis of an important future industry.

The choice of sensor for robot control structure depends on the application. However, a robot vision is the most powerful yet flexible type of environmental feedback which has led to a good deal of R & D effort in this area.

11.1.1. Sensors for use in Robotics

Almost throughout this book, the terms "Sensors" and "Transducers" have been used interchangeably. However, when employed in association with robotic controls, the terms must be understood in a correct perspective.

Since these devices were originally meant as an extension of human senses, it followed that they should be called sensors, but in majority of cases, the tasks were accomplished, rather indirectly by transforming the physical input energy into an electrical output signal which could be measured. This energy transformation or energy transduction process led to sensors' other common name, transducers.

A **transducer** is therefore defined as an **elementary** device capable, within a given field of measurement, of converting a physical input quantity into an electrical output quantity. The transducer itself does not contain any further processing beyond this energy conversion. A sensor is defined as a non-elementary device, usually based on a transducer, capable of converting a physical input quantity into an electrical output quantity, **and** of processing it in accordance with a given algorithm, to provide an output suitable for interfacing to a process control system such as a computer.

The main difference lies therefore in the non elementary nature of the sensor, that is in its capability to embody functions other than the basic energy conversion.

Those sensors which can interact with the control computer to provide data manipulation are termed intelligent sensors while those which can provide the computer with output data only and thus require longer communication times, are called "Non-Intelligent Sensors".

11.1.2. Light Transducers for use with Robots

Light transducers convert electromagnetic energy to electrical energy by absorption. For use with robots, these work in visible spectrum but solid state silicon devices extend the useful range into near IR region as well. A variety of light transducers are made use of, *e.g.*

- **Light dependent resistors (LDR).** These are used mainly in low frequency control applications where their slow response time and relatively high hysteresis are acceptable.
- **Photodiodes and photocells.** A photodiode can be used as a photocell in the absence of an external bias but exhibits a smaller junction capacitance under reverse bias and is therefore suitable for high speed applications such as light pulse measurements, as required in optical range finder systems.
- **Phototransistors.** It is essentially a photodiode-transistor combination. Phototransistors are used mostly in low cost, medium speed and medium sensitivity light detection but are getting replaced by photodiode-operational amplifier combinations.
- **Avalanche photodiodes.** These provide higher output currents then ordinary photodiodes through a cascade multiplication of ions. Being very sensitive high frequency devices, these are suitable for wide dynamic range, broadband light amplification such as is the case in the front end of optical rangefinders.
- **Photo multipliers.** These are vacuum photoemissive devices based on a photocathode, a series of electron multiplying electrodes called dynodes and an anode. In view of its wide dynamic range of operation and high speed characteristic the photomultiplier is ideal for high sensitivity, high speed light amplification, such as in the case of scanning laser rangefinders.

11.1.3. Optical Array Transducers

Most often, composite devices called Array transducers are used in robotic systems and these are made up of a number of photosensitive elements known as photosites. Each array element works on the same principles as the elementary transducers described above. A one dimensional array is called linear array or linear imaging device while a two dimensional array is termed area array or area imaging device, shown schematically in Fig. 11.1.

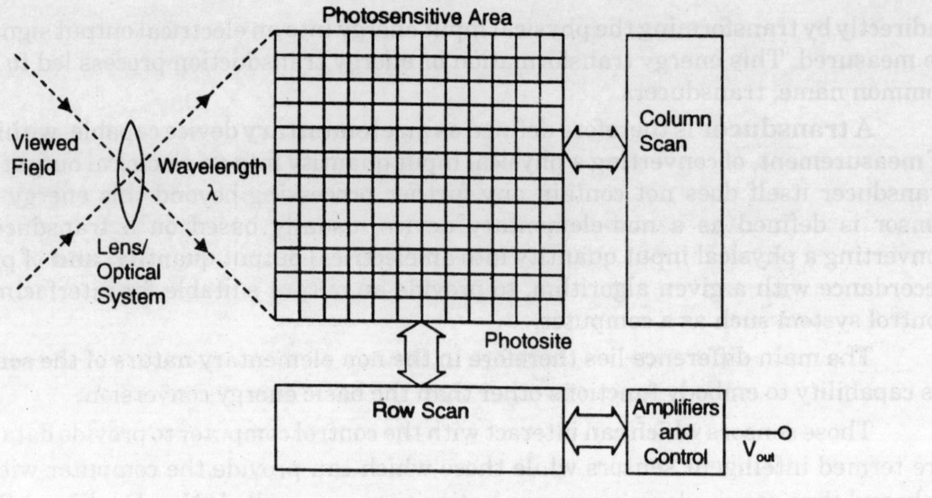

Fig. 11.1. Schematic of optical array transducer.

These array transducers are made up of either vacuum or solid state devices are being adopted for automation and in robotics because of their unique features suited to non contact inspection and measurement.

Vacuum optical array transducers are based on a 2-D photosite array but are not used for linear or 1-D applications. A schematic is shown in Fig. 11.2 which is essentially a basic television camera.

Two types of vacuum TV cameras have become popular. These are called vidicon, which uses a selenium or antimony coated target as a photosensitive area and plumbicon which uses a coating of lead oxide. The scanning of photo sites is achieved via an electron beam (as in case of cathode ray tubes) by using magnetic deflection coils. The principle of operation is illustrated by Fig. 11.3.

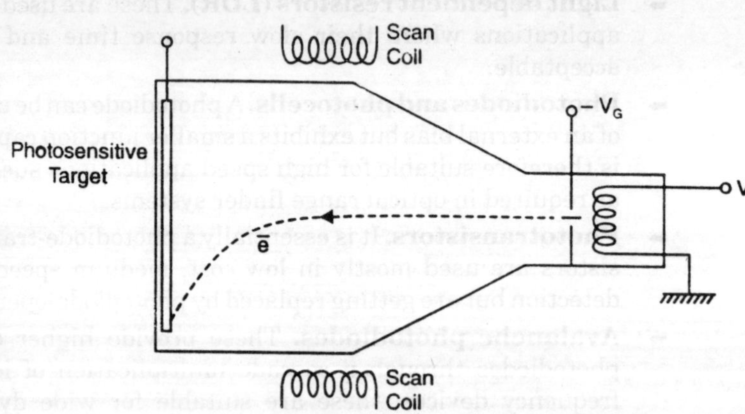

Fig. 11.2. Vacuum optical array transducer (Patterned after TV camer)

The electron beam scans across the target, closes switch S and connects it to the cathode at ground potential. The equivalent capacitor gets charged to a voltage E_1. Then, while the electron beam scans another area of the target and switch S opens, the capacitor discharges through its equivalent parallel component and attains a new lower voltage E_2.

As the electron beam scans across the target area again, it recharges the photosites' capacitors back to E_1 at a rate I_c propotional to the voltage difference (both the capacitor value C_{eq} and time interval Δt remain constant).

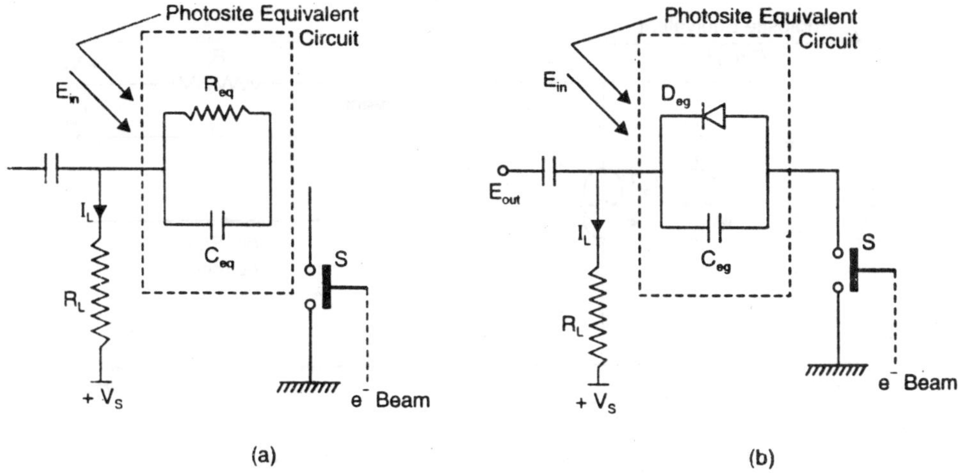

Fig. 11.3. Principle of operation of (*a*) Vidicon and (*b*) Plumbicon.

Therefore, $$I_c = C_{eq}\frac{dE}{dt} = C_{eq}\frac{(E_1 - E_2)}{\Delta t} = K(E_1 - E_2)$$

The value of the parallel component in vidicon is light dependent. The capacitor recharge current I_c is therefore proportional to the incident radiation E_{in} by monitoring the value of I_c during the scanning period, a waveform E_{out} is obtained which is proportional to the illumination levels of the photosites. (*i.e.* to the scene focussed on the target area)

Thus $$I_c = KE_{in} \text{ and } I_L = I_c + I_{dc} = KE_{in} + I_{dc}$$

where I_{dc} is the dark current flowing through the photosite with zero illumination.

The parallel component in case of vidicon is equivalent to a light dependent resistor and its output is therefore non linear.

The parallel component in plumbicon on the other hand, is equivalent to a reverse biased photodiode, which has a linear characteristic.

Vidicon can handle the largest input dynamic range and is therefore less susceptible to image 'Blooming'. It has the lowest unit cost, the lowest sensitivity and a less linear input-output relationship.

By contrast, plumbicon has the lowest image lag, a good sensitivity and a linear input-output response but does exhibit a high unit cost by the necessity to maintain the lead oxide air-free during the manufacturing process.

Both vidicon and plumbicon have a wavelength response similar to that of the human eye.

11.1.4. Solid State Optical Array Transducers

Solid state optical array transducers have been derived mainly from the silicon memory technology but their development has been made possible only because of great strides that microelectronics has made during 1970s and 80s. Solid state cameras, unlike vacuum types, can have either a 2-D photosite array or a 1-D array. These are called area scan cameras in the former case and line scan cameras in the latter case.

A solid state camera based on a photodiode array is described below: The photodiode is a reversed biased *p-n* junction which is light sensitive. The principle of operation is illustrated in Fig. 11.4.

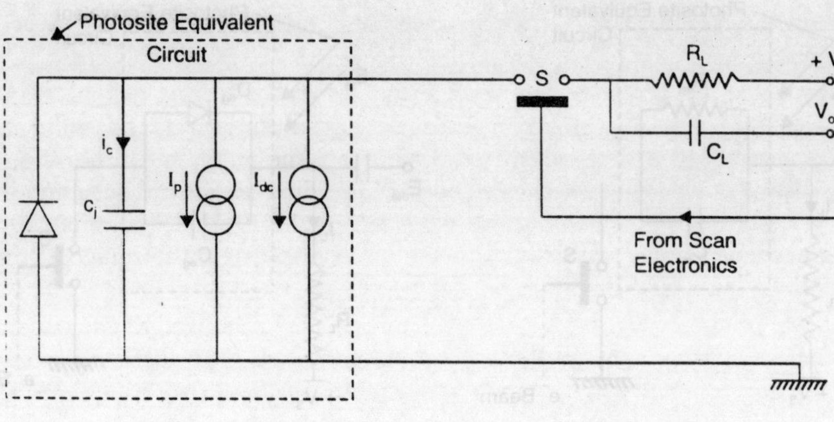

Fig. 11.4. Photodiode array : principle of operation.

When switch S is closed a charge q is stored in the junction capacitance c_j given by the equation :

$$Q = c_j\, v_s = \int i_c\, dt$$

When the switch S is opened, the charge is reduced at a rate proportional to the incident radiation (the photocurrent I_p) and the thermally generated hole-electron pairs (the dark current I_{dc}). For modern, high quality *p-n* junction, $I_p > I_{dc}$. The charge stored within each *p-n* junction is therefore inversely proportional to the incident radiation level. Monitoring the rate of change of this photosites provides a voltage V_{out} proportional to the grey scale image of the scene focussed on the array.

The most common geometry for photodiode arrays in industrial use is what is known as linear array (or linescan) camera. Technological problems involved in making large 2-D photodiode arrays, at a relatively high cost prohibit their use for area scanning.

Line scan cameras offer a simple, low cost method of detecting edge position, which can be used to measure object size and dimensions.

Dynamic Random Access Memories (DRAM)

These optical transducers make use of an ordinary DRAM chip whose ceramic package has had the lid removed so as to allow light to fall on to the memory cells. These devices therefore work by memorizing the light pattern focussed on the memory cells plane. The principle of operation includes three basic steps:

(*i*) The photosites array is coated with electrical charges, all the memory cells are thus filled with logic '1's.

(*ii*) The memory cells are then disconnected from the data bus whilst being exposed to light which, by creating electron-hole pairs, (by photo electric effect) beneath the SiO_2 gates, allows each cell to slowly discharge at a rate dependent on the amount of light falling on it.

(*iii*) After a time interval or exposure time, the memory array is read, thus providing binary data corresponding to the light pattern falling on the array during the exposure time.

The features of this simple, yet very effective area scan camera, depend, on the MOS memory technology used. Memory cell design, in fact, aims to reduce the cell size, the stray capacitances and the signal path lengths in order to obtain the highest cell density with the lowest access time.

This may lead to a physical cell layout unsuitable for use as a light transducer and therefore not all DRAM chips available on the market are suitable for use as a solid state camera. Even those that have been commercially used require some image pre-processing in order to produce an acceptable image, such as the removal of black line caused by the memory cells power supply rail. Two other drawbacks of these devices are the limited dynamic range (and the relatively long exposure times required) and the binary nature of the image. (Since each cell output can only be a logic 1 or 0).

A grey scale image using this device can, however, be produced at the expense of image acquisition time by overlaying several binary images taken with different exposure times. DRAM cameras are inexpensive in comparison with other optical array transducers and as such these are widely accepted in low cost vision systems. A binary vision sensor produces images whose pixel or picture element values are either a black or white luminosity level equivalent to a logic 0 or 1, hence the name "Binary". A block diagram is shown in Fig. 11.5. The complete picture is therefore only a series of logic 1 and 0. This allows easy distinction of dark objects on light background, (and vice versa) and, in view of the low visual data content, fast image data manipulation such as object perimeter and/or area calculations.

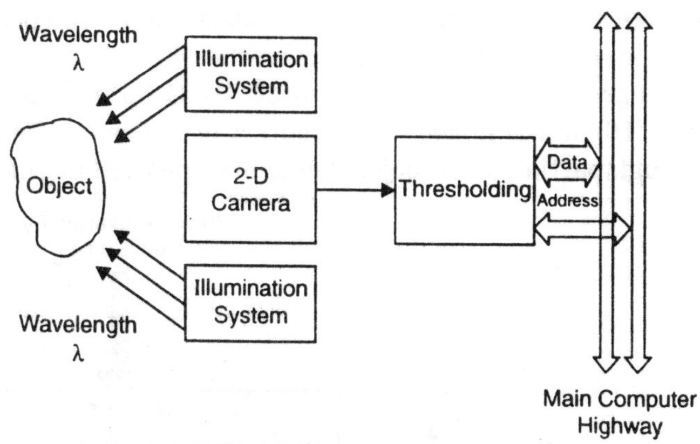

Fig. 11.5. Binary vision sensor (Block diagram).

11.1.5. Scanning Optical Range Finder

In the area of optical 3-D devices, scanning laser range finder has found wide attention during the last two decades and spanned the field of robotics as well as those of aeronautics and military applications.

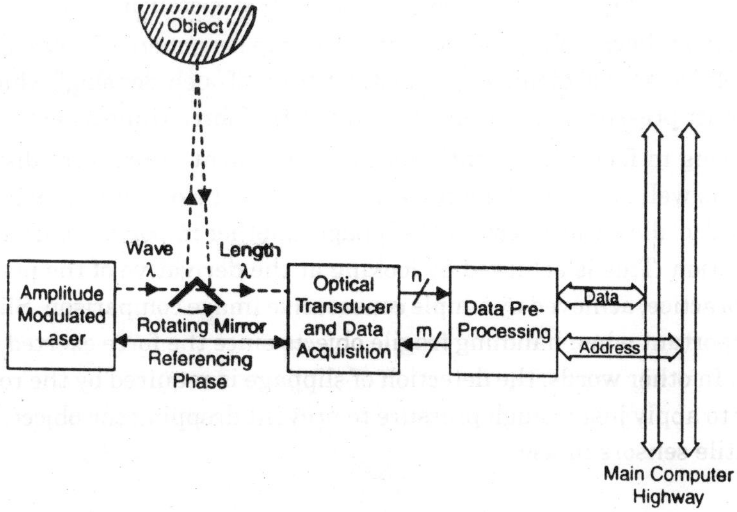

Fig. 11.6. Scanning laser range finder (generalised diagram).

As shown in Fig. 11.6, the technique relies upon mechanical scanning of the target object by a laser beam whose reflected light is processed to obtain the range date ; the resulting depth map has, potentially, a resolution of less than 1 mm. This makes the scanning laser range finder an important 3-D vision sensor but its relatively high cost and inherent mechanical design can be a drawback in applications where low cost and/or rugged sensors are required. This is ought to be one of the reasons why scanning laser range finders have yet to find wide application in the robotics industry.

11.2. TACTILE SENSORS

The limitations suffered by the vision sensors could be overcome by a cooperative use of tactile sensors, a 2-D vision sensor and a range finder. Tactile sensors are based on direct measuring force transducers or the devices that measure mechanical deformation produced by a force acting directly on the transducer itself. The principle is illustrated in Fig. 11.7. Robotic tactile sensors work by measuring the contact pressure between the objective surface and the robot gripper by dividing the contact area into an array of several measurement points (typically 16 × 16), a 2-D image of the contact pressure can be mapped, thereby producing a 3-D view, though partial, of the object being manipulated by the robot gripper.

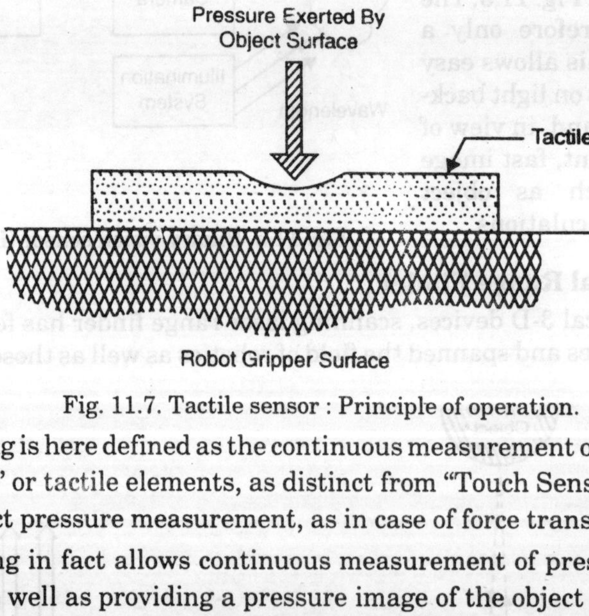

Fig. 11.7. Tactile sensor : Principle of operation.

Tactile sensing is here defined as the continuous measurement of contact pressure within an array of "Tactels" or tactile elements, as distinct from "Touch Sensing" which is essentially a single point contact pressure measurement, as in case of force transducers.

Tactile sensing in fact allows continuous measurement of pressure distribution across the tactels array, as well as providing a pressure image of the object at any point in time; the tactile sensor can also allow the detection of slippage, another important parameter in the field of object manipulation. This is achieved by looking at the derivative of the pressure images, a task which is, in practice, achieved by simple consecutive image comparison. Slippage detection is particularly important when handling fragile objects since the force exerted on their surface must be minimal. In other words, the detection of slippage is required by the robot computer to allow the gripper to apply just enough pressure to prevent dropping the object. There are three main types of tactile sensors in use.

11.2.1. Resistive Tactile Sensors

Resistive tactile sensors are the most popular because of their high sensitivity and *x-y* resolution. These can be based on carbon fibers, conductive rubbers and various specially developed media impragnated with conductive dopants. Although these exhibit the required property of a change in resistance in response to applied pressure, the last category has achieved the maximum popularity because of its lower hysteresis and larger dynamic range.

The construction of resistive tactile sensor is like a two layer piezoresistive strain gauge, having two electrode patterns on separate parallel planes. The electrodes, 1 mm apart are based, as usual, on thin layers of doped silicon or germanium, into rectangular mat sections. The resistance changes are measured either by vertical or top electrodes or horizontal or bottom electrodes (shown in Fig. 11.8).

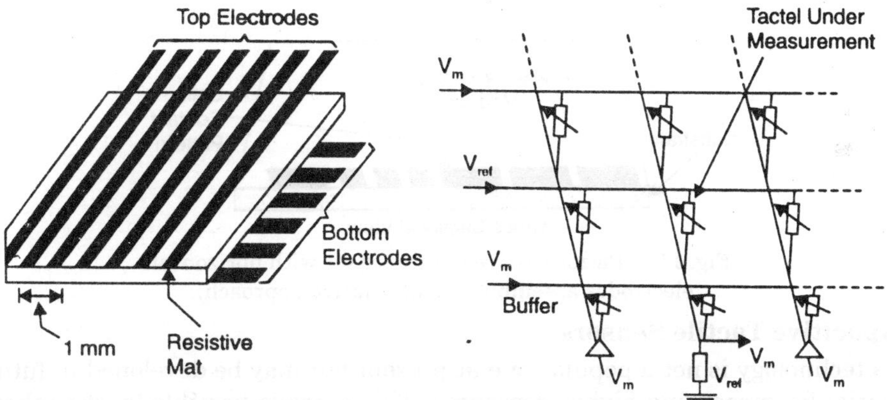

Fig. 11.8. (*a*) Piezoresistive tactile sensor and its interfacing.

The resistance changes are thus measured in line with the applied force vector or perpendicular to it.

The advantage of such a construction is the higher *x-y* tactel resolution achieved by having two electrode patterns on separate parallel planes. However, the life of the top electrode which needs to be deformed mechanically by the applied force for each measurement, becomes shorter.

The sensitivity characteristic curves of such a piezoresistive tactile sensor is shown in Fig. 11.8 (*b*).

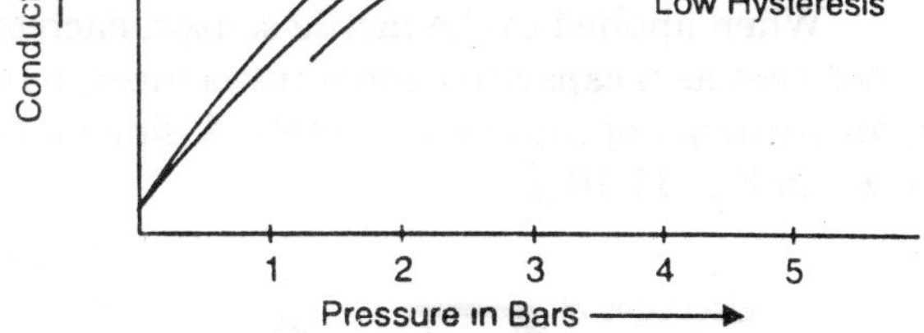

Fig. 11.8. (*b*) Sensitivity characteristic.

An alternative construction approach is shown in Fig. 11.9. Here, resistance change is measured horizontally. The main advantage is the absence of an electrode between the resistive material and the applied force which increases its operational life.

However, accomodating both electrode patterns underneath the resistive medium reduces the *x-y* resolution of the sensor.

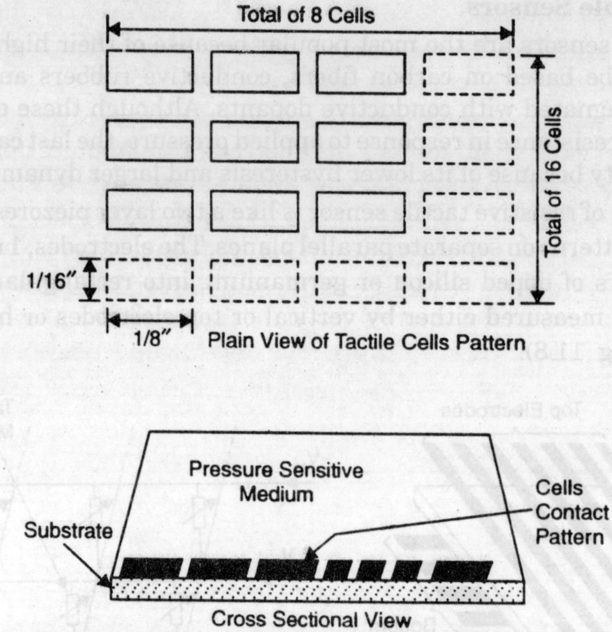

Fig. 11.9. Piezoresistive tactile sensor with horizontal electrodes assembly (an alternative approach).

11.2.2. Capacitive Tactile Sensors

This technology is not a popular one at present but may be developed in future in view of its potential for measuring higher pressures. This is made possible by the inherently less malleable and therefore more mechanically robust construction of capacitive transducers compared with resistive ones. Of-course it is at the expense of a bigger tactel size resulting in a lower x-y resolution.

A capacitive force transducer is shown at Fig. 11.10 (*a*). The change in capacitance is here proportional to the pressure exerted on the plates and depends on the distance d between the plates.

or
$$\frac{\Delta C}{C} = \frac{d}{\Delta d}$$

When applied to the tactile sensor, each tactel behaves as a capacitive force transducer, based on an x-y array of capacitive tactels. A side view is shown in Fig. 11.10 (*b*).

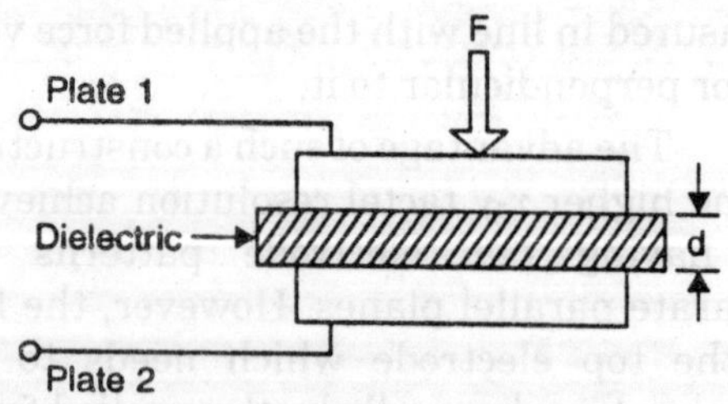

Fig. 11.10. (*a*) Principle of operation of capacitive transducer (compression type).

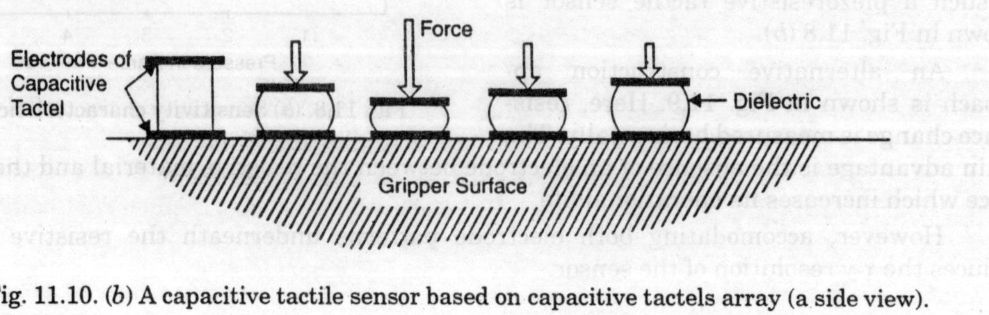

Fig. 11.10. (*b*) A capacitive tactile sensor based on capacitive tactels array (a side view).

11.2.3. Optical Tactile Sensors

Optical tactile sensor relies on changing the light absorption characteristics of the transparent medium under the effect of an applied force.

Several transparent plastics and glass fibers exhibit this property. One way of implementing is based on *x-y* array of light sources and detectors communicating via glass fibers or arranged on the edge of a square sheet of transparent plastic as shown in Fig. 11.11 (*a*).

Another optical technique used to build a tactile array is shown in Fig. 11.11 (*b*) and is based on the slotted optical switch, where the force is applied to a plunger within each individual tactel. This causes partial interruption of the light path between an LED and a photodiode whose output provides a measure of the applied force. A tactile sensor based on this principle is available commercially from M/S Lord Corporation and their agents worldwide.

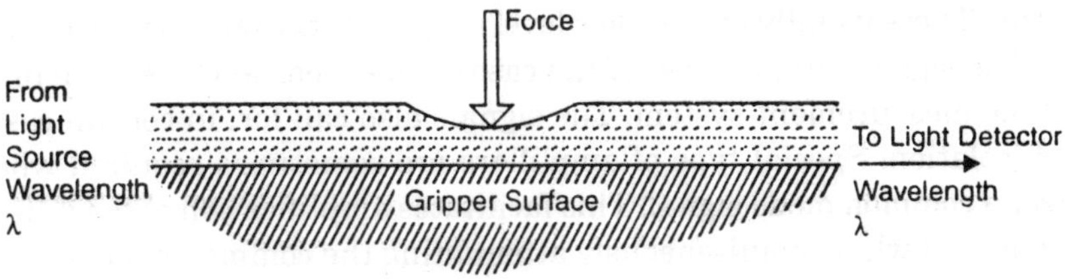

Fig. 11.11. (*a*) Optical tactile sensor : principle of operation.

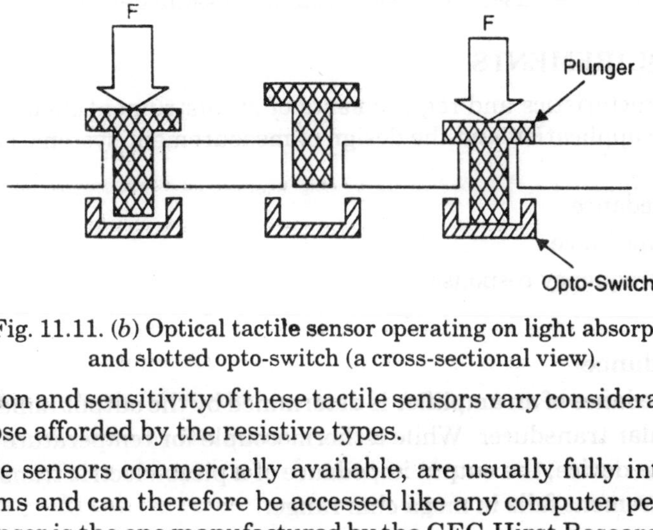

Fig. 11.11. (*b*) Optical tactile sensor operating on light absorption and slotted opto-switch (a cross-sectional view).

The resolution and sensitivity of these tactile sensors vary considerably but, on the whole, are lower than those afforded by the resistive types.

A few tactile sensors commercially available, are usually fully interfaced to one of the popular bus systems and can therefore be accessed like any computer peripheral. An example of such a tactile sensor is the one manufactured by the GEC-Hirst Research Laboratories which already contains and IEEE multibus interface.

BIBLIOGRAPHY

1. S.R. Ruocco, *Robot Sensors and Transducers*, Halsted Press, John Wiley 1987.
2. H.J. Warnecke and R.D. Shchraft, *Industrial Robots : Applications Experience*, I.F.S. Publications England, 1982.
3. J.M. Kauffman and JP Taillard et el, *Robot Technology,* Vol. 4, Prentice Hall Inc., 1983.
4. T.W. Polet, *Industrial Robots : Design, Operation and Application*, Glentop Publishers Ltd. 1985.
5. D. Mccloy and M Harris, *Robotics : An Introduction*, Open University Press, 1986.

12

Transducer Interfaces : Signal Conditioning, Signal Generation And Signal Processing

In general, the signals obtained from transducers are not readily suitable for measurement and control. These usually fall in a low level range of less than a few millivolts and require amplification. The signals are, in most of the cases, non-linear and are to be linearised before connecting to the measurement system. The signals also call for, in certain cases, impedance matching and isolation. Special type of amplifiers are sometimes required when the desired input rides over a common mode signal. This facilitates amplification of the differential signals to an acceptable level while simultaneously attenuating the common mode signals.

We start with the common types of instrumentation amplifiers and their basic characteristics. These are most often used as linear signal processing devices.

12.1. BASIC REQUIREMENTS

The basic characteristics and requirements of an instrumentation amplifier are determined by the specific application and the design of measurement system. These basic requirements are :

- Input Impedance
- Output Impedance
- Gain and frequency response
- Noise

12.1.1. Input Impedance

The input impedance of an amplifier is determined by the output impedance of the signal source or the particular transducer. While a thermocouple for temperature measurement is a low output impedance device, the output impedance of a piezo-electric transducer for acceleration and force measurement falls in mega ohm range.

An ideal amplifier for use in a measurement system is to be of an infinite impedance at input. A non-infinite input impedance introduces an error in the overall system sensitivity. A very high input impedance amplifier can be used to reduce this error. High input impedance amplifiers are also essential to prevent changes in the system sensitivity due to changes in source (or transducer) impedance. For example, in case of an amplifier to be used with a strain gauge bridge, the output resistance may range from 120 to 600 ohms, depending upon the specific bridge circuit used. In such cases, the input impedance should be high enough to reduce the loading errors to a tolerable limit or alternatively, the system sensitivity should be re estimated for every particular type of bridge circuit used.

As a practical approach, the input impedance of an amplifier is recommended to be approximately ten times the output impedance of the transducer source. The resulting signal will be then reduced by not more than 10%, due to loading effect. This is taken to be within permissible limits.

12.1.2. Output Impedance

The output impedance of an amplifier is determined by the input impedance of the sub system that follows ; the analog indicating instrument or a recorder, as the case may be. Again, the loading error is to be below a minimum desirable limit. The amplifier output resistance or impedance is always kept sufficiently low for otherwise a recalibration will be needed alongwith the driven device used.

Suppose an amplifier is required to drive a null galvanometer of resistance 100 ohms. If the deflection sensitivity of the galvanometer is 0.2 cms/mA, the amplifier used should have an output voltage swing of ± 5 V and a current drive capability of ± 50 mA for a deflection span of ± 10 cm. If the loading error is to be kept less than 0.2%, the output resistance of the amplifier should be less than 0.2 ohms.

12.1.3. Gain and Frequency Response

A certain amount of signal attenuation always occurs at the amplifier input which is dependent upon the output impedance of the transducer source and the input impedance of amplifier. The effective gain or amplification is therefore the product of this attenuation and the amplifier gain factor. Since the amplifier gain directly influences the system sensitivity and accuracy, an instrumentation amplifier is required to have a high gain stability.

The frequency response of an amplifier indicates the variation of amplifier gain with frequency. This is specified as a frequency at which – 3 dB loss in gain occurs or a certain percentage change in the gain results. The typical frequency response of amplifiers tends to result in a lesser amplification or gain at higher frequencies. If a deliberate attempt is made to enhance the frequency response, the high frequency response may peak up. In such cases, if G be the gain error tolerable for a system, then the frequency response may be decided by the frequency range over which the deviation in gain is less than ± G% of a specified mid band gain.

The frequency response of an amplifier is also influenced very often by its output impedance. Therefore, when the system response is computed, it is necessary to consider the final load impedance also as a part of the system.

12.1.4. Noise

All the active devices used in amplifiers get influenced in their performance by internally generated noise, which is always frequency dependent. The spectral density of noise is expressed as : $V/\sqrt{f}$ for noise voltage and $A/\sqrt{f}$ for noise current where f denotes the frequency in Hz. The noise spectral density remains constant over a wide range of frequencies, typically from 100 Hz to a few hundred KHz. The noise spectral density increases at very low frequencies due to the presence of flicker noise and also at high frequencies due to the limitations in the amplifier frequency response. This is shown in Fig. 12.1.

The manufacturers of operational amplifiers often specify average value of noise referred to the input over a specified frequency band. For DC applications, peak to peak noise for a frequency range 0.01-10 Hz and for AC applications, the total noise over a 20 Hz-20 KHz frequency range is otten specified.

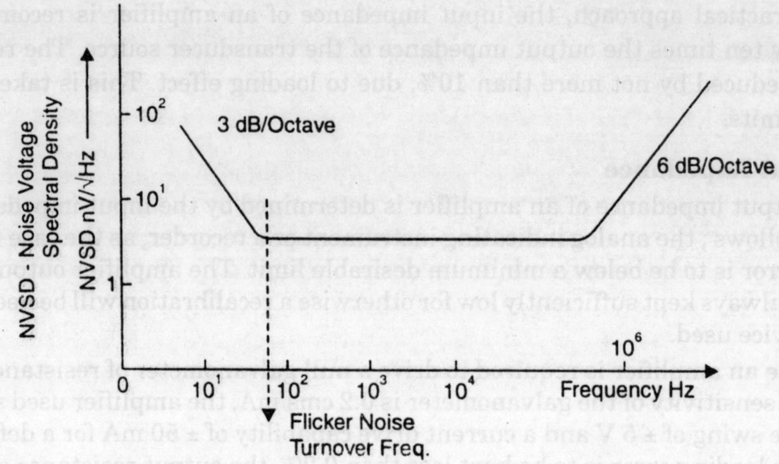

Fig. 12.1. Noise characteristics of amplifiers.

12.2. DC AMPLIFIERS

In practical DC amplifiers, the output may not necessarily have a zero DC value even if the input terminals are shorted. This error is expressed as an average offset voltage defined as the equivalent input voltage required to ensure zero output. This offset voltage is independent of the amplifier gain.

The presence of a finite DC current in the input leads, known as the input bias current, introduces another error in the DC amplifier. This produces an additional offset depending upon on the DC output impedance of the source. This error assume importance when dealing with amplifiers having high input impedance coupled to transducers having high DC output impedance. When the output impedance of the transducer source exceeds a few megaohms, it might become necessary to use dc amplifiers having bias currents of a few nanoamperes or less.

12.3. OPERATIONAL AMPLIFIER

An operational amplifier is basically a DC amplifier with a very high gain, of the order of 10^4 to 10^8. Such a device can perform summation and integration and can also be used as a comparator with suitable feedback networks.

12.4. DIFFERENTIAL AMPLIFIER

A differential amplifier is shown in Fig. 12.2. The basic circuit consists of two matched transistors working under the same operating conditions. Ideally, the amplifier responds to

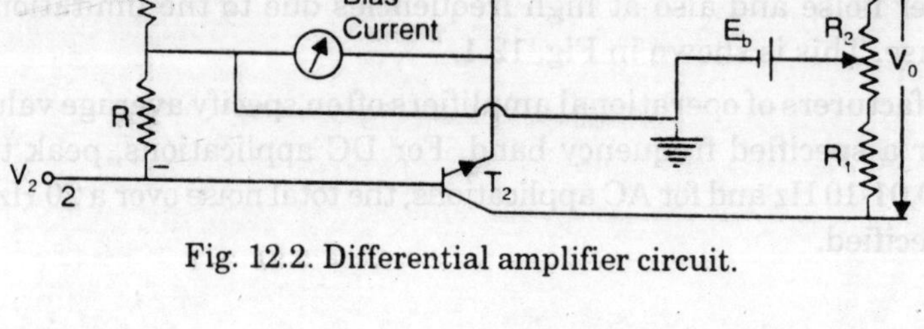

Fig. 12.2. Differential amplifier circuit.

difference signals between the input terminals and ignores interference signals that appear in phase at both the terminals.

The differential input $V_i = V_1 - V_2$ and output $V_0 = A_d (V_1 - V_2)$ where A_d denotes differential gain.

When the input terminals 1 and 2 are kept floating and no input signal connected, V_0 should be zero but due to the presence of stray electric fields, V_1 and V_2 may assume the same potential V_{cm} because of their proximity. If two halves of the amplifier are not matched perfectly, a voltage will appear at the output terminals :

$$V_0 = A_{cm} \cdot V_{cm}$$

where A_{cm} denotes common mode gain and V_{cm} the common mode voltage. The ratio A_d/A_{cm} is called common mode rejection ratio or CMMR, which represents the imperfect functioning of the differential amplifier.

This drawback is overcome by instrumentation amplifier.

12.5. INSTRUMENTATION AMPLIFIER

This is basically a committed configuration consisting of operational amplifiers intended primarily for amplification of voltage signals derived from transducers with accurately adjusted gain values. Any variation in gain affects the accuracy of the measured primary quantity. The amplifier is required to amplify low level signals superimposed on the common mode voltage. The significant features are :

- Low drift
- High input impedance
- High Linearity
- High CMRR
- High noise rejection capability
- Low output impedance

The instrumentation amplifier is therefore an improved version of the basic differential amplifier. It differs from the ordinary operational amplifier in the following respects :

- It is a package of operational amplifiers with an accurate and stable feedback to give a desired gain, which is selectable to a precise value by a single external resistance.
- It is a closed loop configuration. The gain accuracy and stability as well as drift are normally specified by the manufacturer.
- It will yield high common mode rejection even with source impedances exceeding 1 MΩ and with source unbalance of a large value (1 to 10 KΩ). On the other hand, a DC amplifier has often a common mode rejection ratio specified for it which is applicable for very low source impedances only.

A unique three operational amplifier configuration employed to hook up an instrumentation amplifier is shown in fig. 12.3. A high CMRR is realised by connecting two non inverting links with operational amplifiers OA_1 and OA_2 alongwith a common feedback resistors R_1.

With voltage applied to one terminal and the other one grounded,

$$V_{01} = +V_1\left(1 + \frac{R_2}{R_1}\right)$$

$$V_{02} = -V_1\left(\frac{R_2}{R_1}\right)$$

Input voltage V_1 is applied at the terminal 1 and terminal 2 is grounded and $$V_{01} = -V_2\left(\frac{R_2}{R_1}\right)$$ $$V_{02} = +V_2\left(1+\frac{R_2}{R_1}\right)$$ Input voltage V_2 is applied at the terminal 2 and terminal 1 is grounded

Therefore, the net effect of applying V_1 and V_2 is to give

$$V_{01} - V_{02} = \left(1+\frac{2R_2}{R_1}\right)(V_1 - V_2)$$

The second stage of the instrumentation amplifier is a differential amplifier with a gain R_4/R_3.

Hence, the net gain $$A = \frac{(1+2R_2/R_1)}{R_4/R_3}$$

The required gain values of A can be obtained by adjusting the feedback resistor R_1 alone. Since the drift errors of the second stage will add to the product of the drift errors of the first amplifier and the gain of the first stage, it is necessary to make the first stage gain to an adequate value. This will prevent to overall drift performance from being significantly affected by the drift in the second stage.

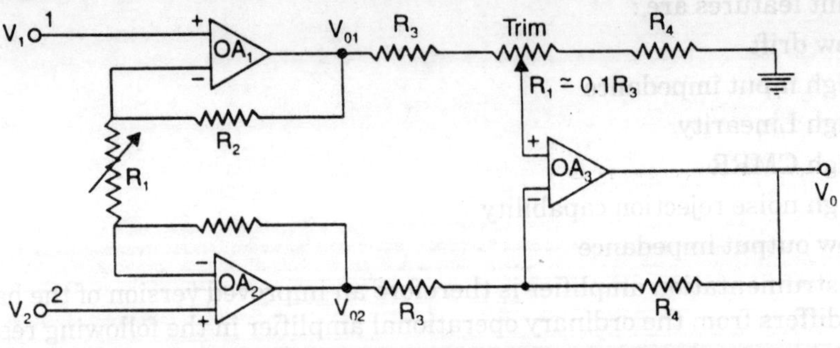

Fig. 12.3. Instrumentation amplifier configuration.

A trimmer $R_1 = 0.1\ R_3$ is included to improve CMRR of the second stage. Thus, there is an overall improvement in the CMRR by a factor $(1 + 2R_2/R_1)$. The dual input stage OA_1 & OA_2 is seen to provide a differential gain of $(1 + 2R_2/R_1)$, while having a common mode gain of unity only.

Typical values of CMRR range from 80 dB to 110 dB. The dedicated versions of the instrumentation amplifier for use with DC Wheatstone bridge networks possess high CMRR values upto 140 dB for 0-50 Hz voltages. Use of precision metal film resistors ensures a high gain accuracy and because of a large negative feedback employed, the amplifier has a good linearity, typically about 0.01% for a gain of less than 10^4. The output impedance is low, in milliohm range.

The input bias current is determined by those of the input amplifiers OA_1 and OA_2. The use of operational amplifiers μA725 C for OA_1 and OA_2 gives an input bias current of about 42 nA.

12.5.1. Amplification of Signal from a Strain Gauge Bridge

Let a strain gauge wheatstone bridge be connected to a differential amplifier shown in Fig. 12.4. The equivalent voltages and resistances from the inverting and non inverting

terminals to the ground will normally be unequal. The equivalent voltages will be equal only when the bridge is balanced perfectly. Even at balance, the resistance values are unlikely to be equal in which case, the difference will be considered as source unbalance resistance and there will be an error in output voltage V_0 due to the common mode signals at the input terminals.

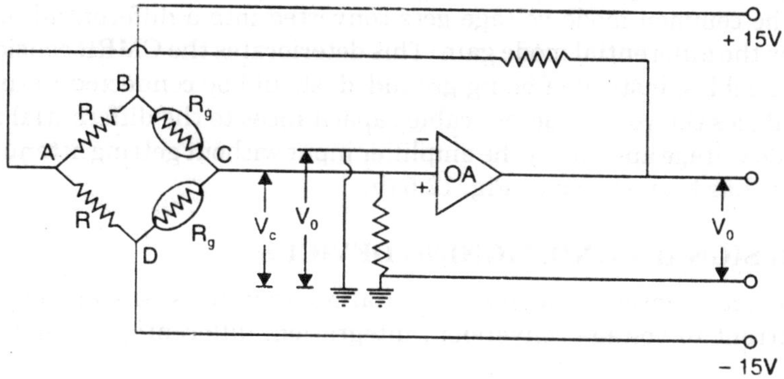

Fig. 12.4. Strain gauge wheatstone bridge with differential amplifier.

The higher the voltage EBD, larger will be the error. Hence, for such situations, where very low level voltages present alongwith high common mode voltages are required to be amplified, an instrumentation amplifier will be the only choice.

12.5.2. Guarding of Input Against Cable Capacitance

Since the instrumentation amplifier has a very high CMRR, its gain for common mode signals appearing at its input terminals is very low as compared to its gain for differential mode

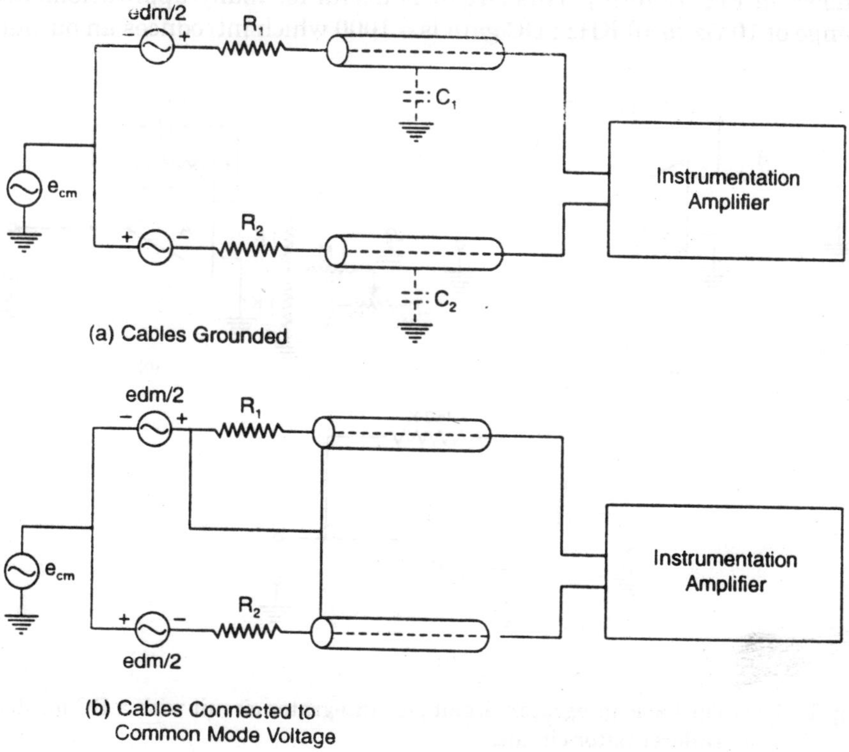

Fig. 12.5. Guarding against cable capacitances.

signals. A one metre length of cable is usually run from transducer to amplifier. Because of the mismatch in the capacitance of this cable length, the common mode signal at the amplifier will be different from what existed at the transducer. In the equivalent circuit shown in Fig. 12.5 (*a*), V_{cm} is the common mode voltage at the source and V_{dm} is the differential mode voltage. If R_1C_1 and R_2C_2 are not equal (where R_1, R_2 are source resistances and C_1, C_2 the cable capacitances), the common mode voltage gets converted into a differential mode voltage and gets amplified by the differential mode gain. This deteriorates the CMRR considerably. In order to undo this, the cables, instead of being grounded, should be connected to the common mode voltage. This reduces the voltage across cable capacitances to the differential mode signal and the common mode voltage appears at the amplifier input without getting attenuated by the cable capacitance. This has been shown at Fig. 12.5 (*b*).

12.6. OTHER SIGNAL CONDITIONING DEVICES

Besides using differential or instrumentation amplifiers, signal conditioning involves attenuation, current to voltage conversion, integration, differentiation and impedance level transformation.

12.6.1. Integrators and Differentiators

Integrators are used when the signal from an acceleration transducer has to be converted to velocity or when a velocity signal is to be converted into displacement. Conversely, to obtain velocity from displacement or acceleration from velocity, differentiators are required. The basic integrator circuit using an operational amplifier is shown at Fig. 12.6 (*a*) but has little practical use since the non zero offset errors get amplified by the dc gain of the operational amplifier and produce high output offset errors. These errors can be reduced by lowering the DC gain by a DC feedback [shown in Fig. 12.6 (*b*)]. This circuit is useful for many applications over a limited frequency range of 10 Hz to 10 KHz ; DC gain is – 1000 which introduces an output drift of less

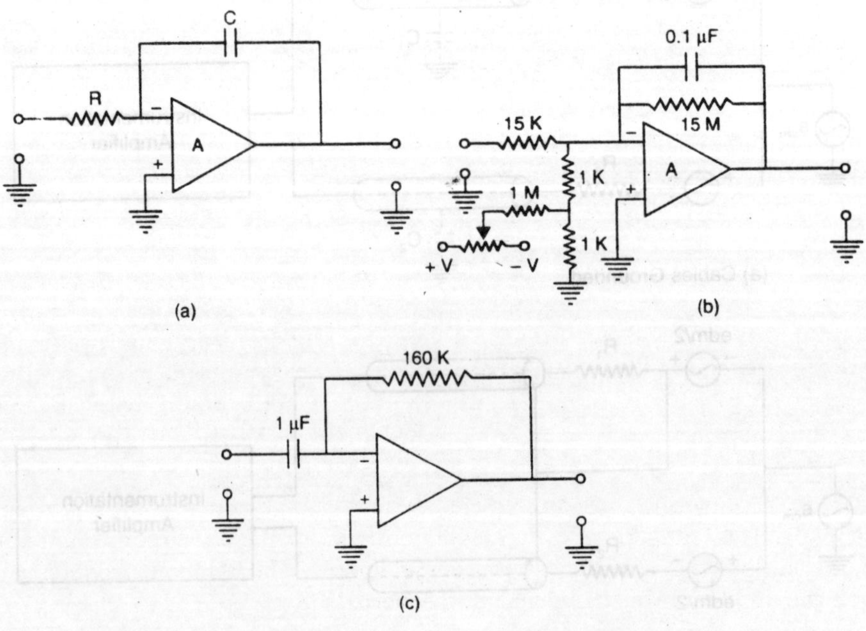

Fig. 12.6. (*a*) The basic integrator circuit ; (*b*) Integrator circuit with a DC feedback ; (*c*) Differentiator circuit.

than a few millivolts. To ensure that the output at 10 KHz does not become comparable with the drift, the input at this frequency should be of the order of a few volts. The useful operating range of this integrator can be shifted to a frequency range of 1 Hz to 1 KHz by increasing the feedback capacitance to 1 μF.

A differentiator can be obtained by the arrangement shown at Fig. 12.6 (*c*). A limitation is that due to roll-off of the amplifier gain at high frequencies and the differentiating characteristics at the operating frequency, the close loop gain gives a peaky response as a result of which, the differentiator become very noisy and amplifies its own internal noise as well by as much as 10^5 times in the 1 KHz range. A modified differentiator, useful in practical situations, uses a μA748 operational amplifier. A resistance put in series with the input circuit capacitance flattens the response at a frequency around 30 KHz.

12.7. SIGNAL GENERATION AND SIGNAL PROCESSING

A complete measurement system very often requires some form of excitation in a majority of cases (like excitation of certain passive transducers) and one or more types of processing circuits to convert the outputs of transducers into useful data. The excitation may be a constant DC voltage or current or a stabilised AC signal. In case of swept measurements, the need arises for varying the frequency of the excitation source or amplitude in a predetermined manner with time. When the transducer characteristics are non-linear, it may be advantageous to linearize the output. Some basic types of signal generators and processors often used in instrumentation systems and discussed below.

12.7.1. Oscillators and Signal Generators

Certain transducers and bridge networks with one or more arms consisting of transducers are excited by AC supply voltages, mostly of sinusoidal waveform.

The amplitude and frequency of the voltage in all such cases should be stabilised to a high degree of accuracy. The frequency may be chosen at any value within the range of 0.001 Hz to 1 MHz, depending on the actual application. These are treated as fixed frequency oscillators.

When dynamic behaviour of physical systems has to be experimentally determined, signal generators whose frequencies can be varied over a desired range, are used. Waveform could be sinusoidal, square wave or saw tooth form but amplitude should be held constant while frequency is varied.

Voltage to frequency conversion has a particularly important role in A-D convertors and other data acquisition systems. The circuits performing this conversion are known as voltage controlled oscillators and are used in conjunction with a transducer, usually a resistive type, to generate output voltages of a frequency proportional to the amplitude of the input voltage. The output signals thus brought out into frequency or time domain enable counting operations for a digital readout as also transmission of signal to a remote location, without any error due to transmitting link.

Similarly, tuned oscillators enable conversions of a measurand into frequency when an inductive or capacitive transducer forms an element of the tuned LC circuit. Their output signals are used for modulation of a carrier signal for telemetry. The resonant frequency of LC tuned circuit when the amplitude of the measurand is zero, is given by

$$f_0 = \frac{1}{2\pi\sqrt{LC}}$$

Variation of L or C occurs in such a way that the fractional changes in them $\Delta L/L$ or $\Delta C/C$, are proportional to the instantaneous amplitude of the measurand and hence the instantaneous value of frequency deviation Δf is directly proportional to the fractional change, provided that $\Delta C/C$ is < 0.05.

The frequency deviation is so approximated that

$$\Delta f = f_0 \left(1 \pm \frac{\Delta C}{C}\right) = f_0 \left(1 \pm \frac{\Delta L}{L}\right).$$

12.7.2. Crystal Oscillators

These use quartz crystal, have a high frequency range of 100 KHz-10 MHz and frequency stability of 10^{-6} to 10^{-10}. These play an important role in instrumentation and are often employed as accurate time references.

The circuit used when a nearly sinusoidal output is desired, is shown in Fig. 12.7. The crystal constants and the value of C determine the frequency of oscillation.

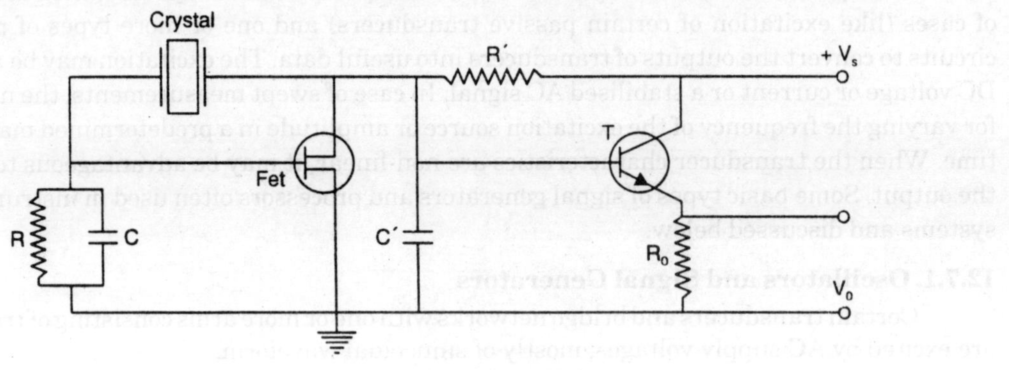

Fig. 12.7. Crystal oscillator.

12.7.3. Pulse Generator

Pulse generators and square wave generators differ in respect of the mean level of pulse train. The square wave generator has the mean level of zero and positive and negative pulses are each of duration equal to T/2 where T is the period of the wave. In a pulse generator, pulse is on for a time T_1 and off for a time T_2, and pulse rate = $1/(T_1 + T_2)$. A simple pulse generator using a UJT is shown in Fig. 12.8, alongwith the wave shape.

The capacitor C gets charged from Vs through the variable resistor R. The charging time can be made to vary by varying R. The charging time can be made to vary by varying R. When

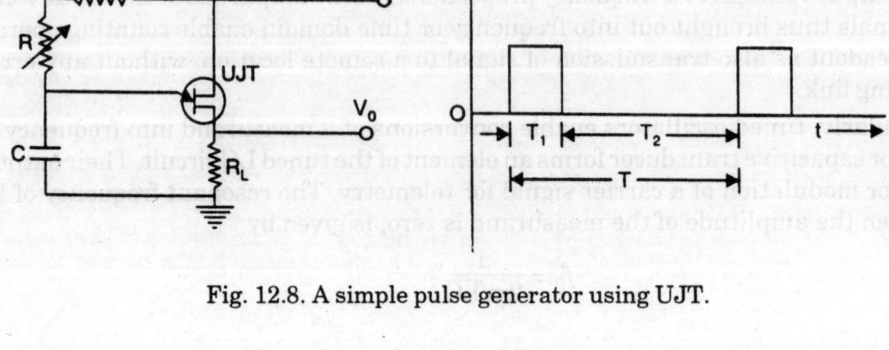

Fig. 12.8. A simple pulse generator using UJT.

the voltage across capacitor is equal to the threshold voltage of UJT, the transistor will conduct and partially discharge the capacitor. The output voltage across R_L is processed suitably to constitute the pulse train desired.

The pulse rate is proportional to 1/RC.

12.7.4. Square Wave Generator

An astable or free running multivibrator using an RC circuit and operational amplifier are the most commonly used forms.

A single operational amplifier performs the task of a level comparator and control of charging/discharging currents.

The feedback circuit R in association with the input capacitor C forms the integrator and the positive feedback through R_1 and R_2 sets the comparator levels. Frequency variation is brought by varying R.

Frequency of oscillation ≈ 1/RC.

Such circuits [shown in Fig. 12.9 (*a*)] are useful upto a frequency of 30 KHz.

Schmitt Trigger is a comparator for which the positive and negative transitions occur at different levels of input voltage. This circuit is shown in Fig. 12.9 (*b*).

The frequency of square wave output voltage is the same as that of the input. Amplitude of the square wave remains constant inspite of fluctuations in the input voltage due to noise.

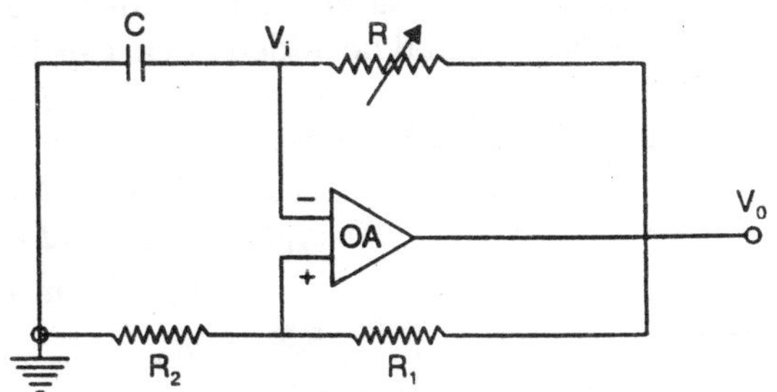

Fig. 12.9. (*a*) An astable multivibrator used as a square wave generator.

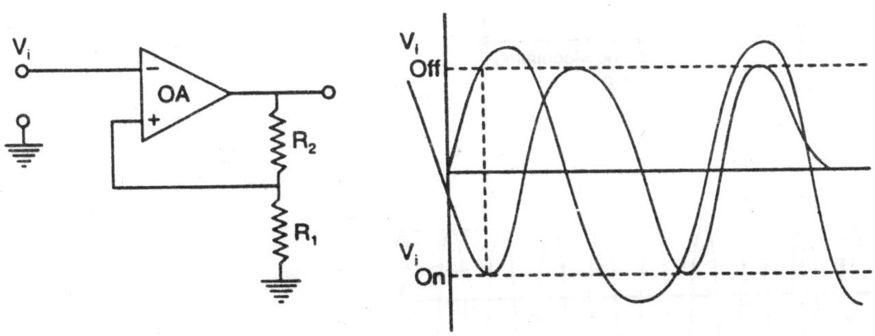

Fig. 12.9. (*b*) A schmitt trigger comparator used as a square wave generator.

12.8. MODULATION

The signal as obtained from transducers is not always amenable for transmission to a remote location either by wire link or radio link. The signal is required to be processed before transmission. The technique is known as modulation in which one of the characteristics of the wave, such as amplitude, phase or frequency is modified by the signal so that it is fit for transmission by the channel. For this purpose, a standard waveform with a specific amplitude, frequency or phase is chosen which is called 'carrier' and signal to be transmitted as 'modulating signal'. The carrier is supposed to be carried without any distortion or loss of information. At the receiving end, the "modulating signal" is processed again and recovered by removing the carrier. This process is called demodulation. The following techniques are used primarily for non linear processing of analog signals.

12.8.1. Amplitude Modulation

The simplest form of amplitude modulation is effected by multiplication of carrier wave signal $e_c(t)$ and the modulating wave signal $e_m(t)$, both periodic and sinusoidal :

If $$e_c(t) = E_c \cos \omega_c t \quad \text{and} \quad m(t) = E_m \cos \omega_m t$$

Then the modulated signal is obtained as :

$$e_m(t) = e_c m(t) = (E_c \cos \omega_c t)(E_m \cos \omega_m t)$$

$$= \frac{E_c E_m}{2} [\cos (\omega_c + \omega_m)t + \cos (\omega_c - \omega_m t)]$$

$(\omega_c \pm \omega_m)$ are the side band frequencies and it is known as double side band suppressed carrier (DSBSC) signal since no carrier wave frequency is present in the modulated wave signal. The modulating signal is contained in two side bands and can be later recovered through demodulation. Let a voltage signal be added at carrier frequency to the DSBSC.

Then $$e_m(t) = E_c \cos \omega_c t + K m(t) E_c \cos \omega_c t$$

But since $m(t) = E_m \cos \omega_m t$, therefore

$$e_m(t) = E_c \cos \omega_c t + KE_m \cos \omega_m t E_c \cos \omega_c t$$

or $$e_m(t) = E_c \cos \omega_c t\,(1 + KE_m \cos \omega_m t)$$

$$= E_c\,(1 + m_a \cos \omega_m t) \cos \omega_c t,$$

putting $$KE_m = m_a$$

where m_a is the modulation index and is controlled to assume various values from zero to unity. Fig. 12.10 (*a*) shows the modulated wave signal. The carrier frequency signal varies in amplitude from zero for $m_a = 0$ to twice its peak value for $m_a = 1$.

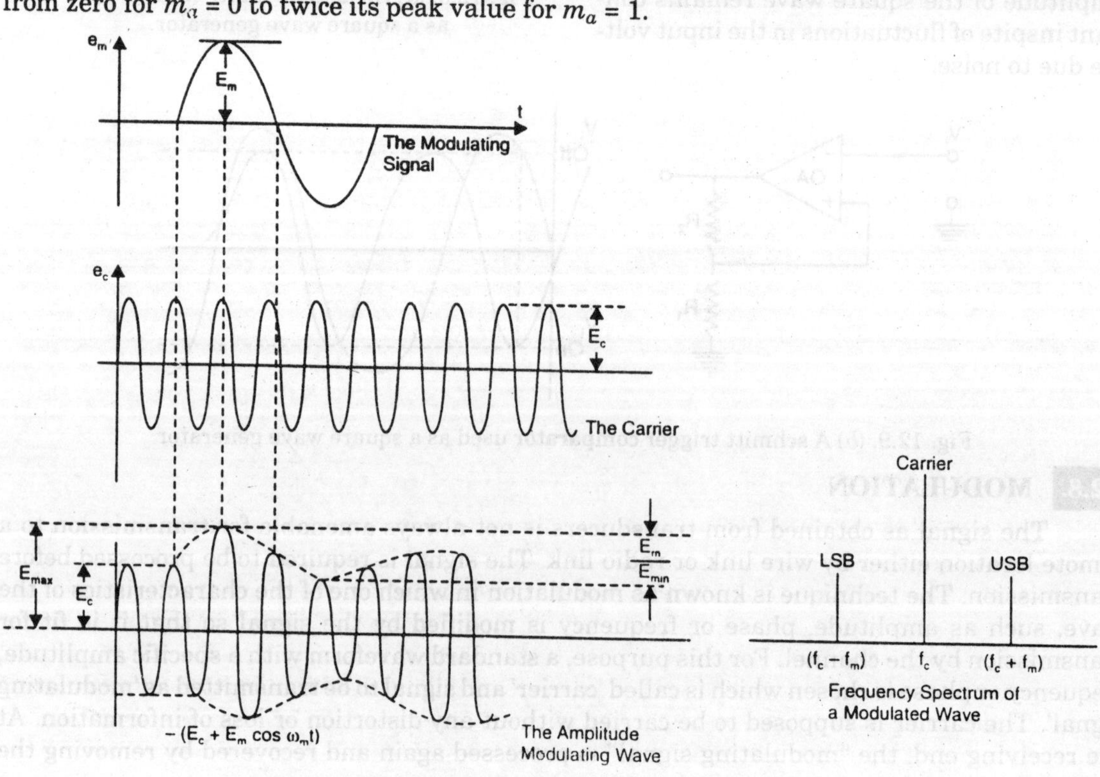

Fig. 12.10. (*a*) The amplitude modulation.

Almost 2/3rd of total power is consumed for transmitting carrier signal only and the rest 1/3rd is available in upper and lower side bands. However, the presence of carrier signal at the receiving end results in proper demodulation. Still, amplitude modulation remains an inefficient

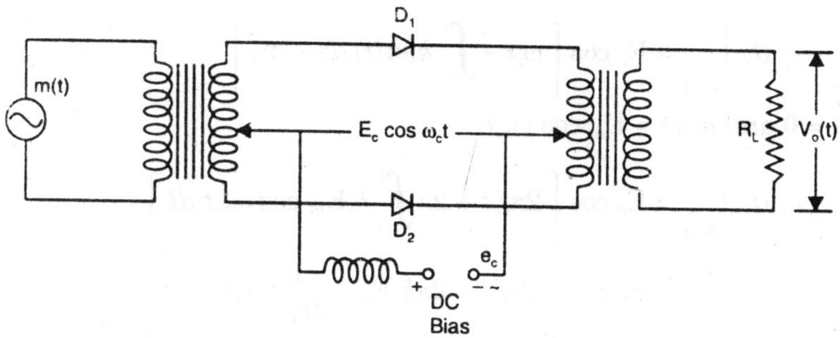

Fig. 12.10. (*b*) The balanced amplitude modulator.

technique of modulation and is rarely adopted since the transmitted signal may get sufficiently curbed in its amplitude due to noise interference and the signal to noise ratio may assume unacceptable levels. A basic amplitude modulation arrangement has been shown in Fig. 12.10 (*b*) and a simple demodulation, in Fig. 12.10 (*c*), wherein the demodulation or rectified voltage appears across the parallel RC circuit and it contains the envelope of amplitude modulated signal wave shown in Fig. 12.10 (*a*).

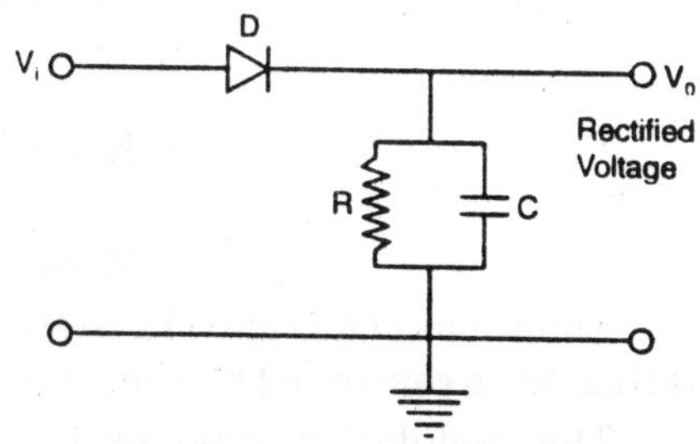

Fig. 12.10. (*c*) Demodulator.

12.8.2. Frequency Modulation

If the unmodulated carrier signal is represented by: $E_c = \cos(\omega_c t \pm \Phi_c)$, (where E_c denotes the peak amplitude), it is called phase modulation if Φ_c is made to vary instant by instant, in proportion to $m(t)$, the amplitude of the modulating signal.

On the contrary, if angular frequency of carrier ω_c is made to swing, instant by instant, in proportion to $m(t)$, it is called frequency modulation.

However ω_t at any instant is given by :

$$\omega_t = \frac{d\Phi_c(t)}{dt} = \omega_c + k_f m(t) \qquad ...(12.1)$$

where k_f is a constant for frequency modulated signals. The carrier signal is of frequency $\omega_c(t)$ in the absence of modulating signal $m(t)$. Thus it is possible to modulate the angle of a carrier signal, while holding its amplitude E_c, constant.

The total angle $\theta_c(t)$ of the modulated wave is obtained from the above by integration, or

$$\theta_c(t) = \omega_c t + \int_0^t k_f m(t) dt + \Phi_0 \qquad ...(12.2)$$

where Φ_0 is a constant of integration representing phase modulated carrier wave at $t = 0$ and it can be treated as zero for frequency modulated signals.

The equation (12.2) represents phase modulation where the modulating signal is given by

$$\int_0^t m(t)\, dt \text{ in place of } m(t).$$

Therefore, a frequency modulated signal can be produced by first integrating the modulating signal $m(t)$ and then using it as input to a phase modulating circuit.

Using equation (12.2), the expression for frequency modulated signal becomes :

$$e_m(t)\Big|_{FM} = E_c \cos\left[\omega_c t + \int_0^t k_f m(t)\,dt + \Phi_0\right]$$

Taking $\Phi_0 = 0$ and $m(t) = E_m \cos \omega_m t$

$$e_m(t)\Big|_{FM} = E_c \cos\left[2\pi f_c t + 2\pi \int_0^t k_f E_m \cos \omega_m t\,dt\right] \quad ...(12.3)$$

$$= E_c \cos\left[2\pi f_c t + 2\pi k_f E_m \cdot \frac{1}{2\pi f_m} \sin \omega_m t\right]$$

$$= E_c \cos\left[2\pi f_c t + \frac{k_f E_m}{f_m} \sin \omega_m t\right] \quad ...(12.4)$$

$$= E_c \cos\left[2\pi f_c t + \frac{\Delta f_c}{f_m} \sin \omega_m t\right] \quad ...(12.5)$$

$$= E_c \cos\left[2\pi f_c t + m_f \sin \omega_m t\right] \quad ...(12.6)$$

Equations (12.1) and (12.5) show that frequency of carrier signal deviates from f_c by $\pm \Delta f_c$ and that Δf_c is controlled by the amplitude of the modulating signal.

The modulation index m_f in eqn. (12.6) is defined as the ratio of deviation of carrier frequency to the frequency of the modulating signal and is given in radians. Its value is governed both by amplitude and frequency of the modulating signal. Treating m_f as peak phase displacement of the carrier signal due to the modulation process, it is seen that the smaller the value of f_m, the larger the phase displacement.

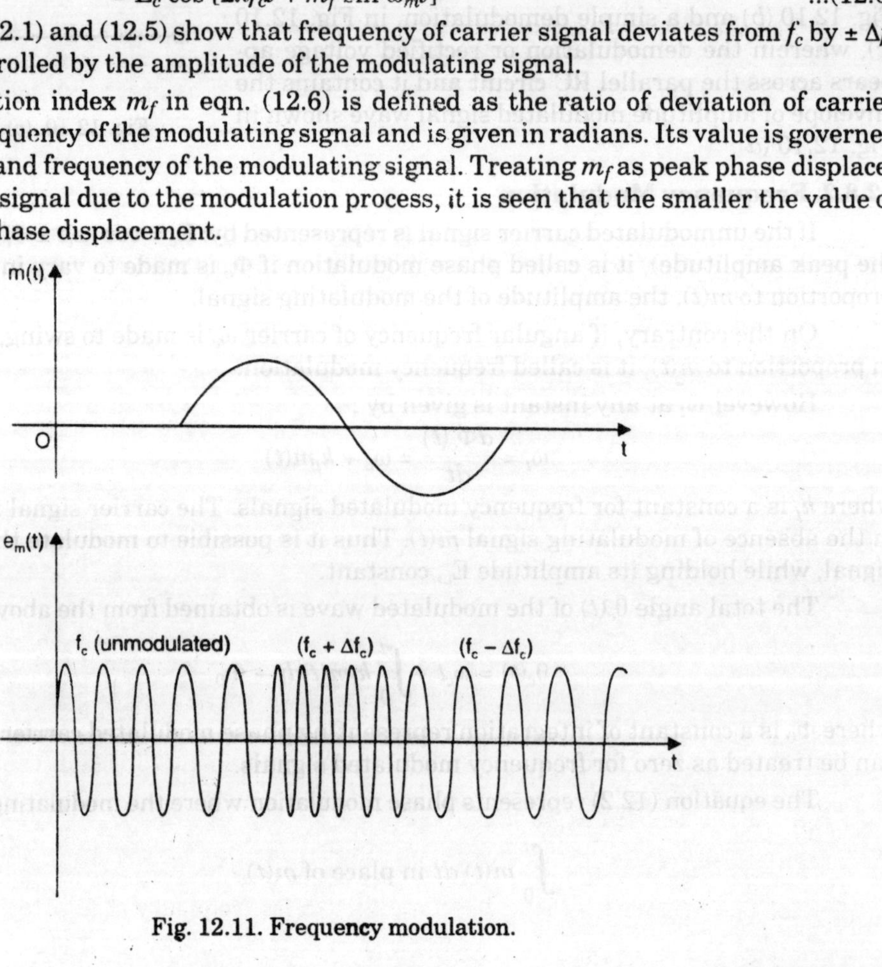

Fig. 12.11. Frequency modulation.

$$\text{Eqn. (12.6) yields } e_m(t)\Big|_{FM} = E_c\,[\cos \omega_c t \cos (m_f \sin \omega_m t) - \sin \omega_c t \sin (m_f \sin \omega_m t)] \quad ...(12.7)$$

Therefore, for modulating signals with f_m much smaller than f_c, the modulated carrier signal can be regarded as a succession of sinusoidal waves of slowly varying period (or frequency) but of constant amplitude as shown at Fig. 12.11.

The generator that produces a carrier of frequency upto 500 MHz, is in many cases, an LC tuned circuit. It provides a sinusoidal waveform for the FM signal ; the modulating signal *m(t)* is usually applied in such a way as to vary the C or L of the LC tank circuit, but proportional to the instantaneous amplitude of *m(t)*.

12.8.3. Direct FM Generators (or Transducer Oscillators)

A proper circuit for frequency modulation may have to be chosen, depending upon whether the data transmitted is obtained from transducers and measuring systems or from microphones etc. Analog transducers whose outputs effect changes in R, L or C, can be made to alter the frequency of oscillation of electronic oscillators, thereby converting signals from analog domain to frequency or time domain. This is an indirect conversion of analog signal to digital. Apart from L or C becoming a part of a transducer system, the L or C may also by made to vary electrically or electronically. The technique is used when the modulating signal is a voltage or current. The self inductance of a saturable core reactor can be varied by changing the dc biasing current through the bias winding and thereby changing the core permeability. Similarly, the dc biasing voltage of a voltage variable capacitor, can be varied, thus bringing about a variation in its capacitance. Semiconductor diodes when operated with a reverse bias, have characteristics suitable for use as voltage variable capacitors. A voltage variable capacitor, commonly known as Varicap or a varactor diode, is kept in parallel with a fixed capacitor C_o of the tank circuit. The capacitance C_v of the varicap changes in proportion to the amplitude of the modulating voltage *m(t)*, and hence causes a corresponding variation in oscillator frequency. Fig. 12.12 shows the circuit.

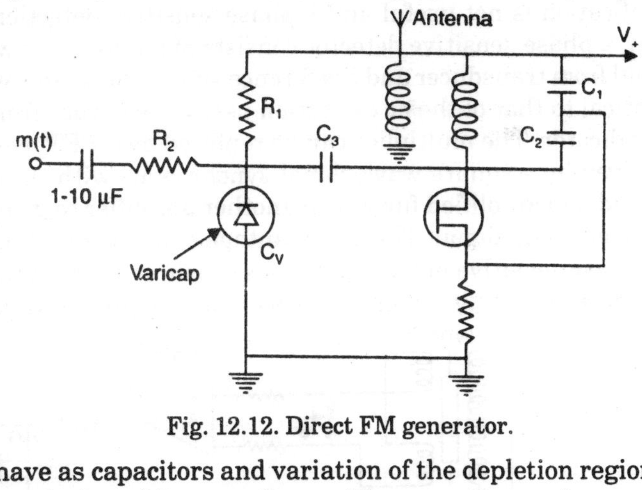

Fig. 12.12. Direct FM generator.

At high frequencies, diodes behave as capacitors and variation of the depletion region of the diode with voltage results in the variation of its capacitance. The capacitance change per unit volume is highest at the lower values of DC bias voltages and the relationship is non linear. Usually, amplitude of the modulating voltage is so chosen that the diode operation is always within the non-conducting region. The capacitance of the diodes is in the region of 10—100 pF at – 4 V bias, and the modulating voltage is coupled to the diode as shown. The coupling of the diode with the tank circuit should in no case, result in the loading of the tank circuit by the source of *m(t)* or vice versa. For centre frequencies of about 100 MHz, the circuit forms part of a colpitt's oscillator.

Drawbacks

Carrier signal frequency drifts when *m(t)* = 0 or in the event of no modulation. The FM generator has to possess a long term stability in its frequency of generation, especially when used for transmission of the measured data. When frequency modulators are used for the first stage of modulation, for telemetry of the measured data, they are commonly known as sub carrier oscillators. The required degree of constancy of frequency should also be ensured; otherwise a continuous check and adjustment of the same has to be undertaken.

The indirect method of FM signal generation provides a constant frequency of the unmodulated carrier. This is made possible by using a crystal oscillator (described in 12.7.2 and Fig. 12.8) as the basic carrier signal generator.

12.9. SIGNAL PROCESSING CIRCUITS

In many measurement systems, the transducer outputs are required to be processed before being displayed or converted into other forms for transmission or control. For example, LVDT excited by a carrier wave, where the signal output has to be processed so that the information can be retrieved faithfully. For data that are transient in nature, quick but temporary storage may become essential (*e.g.* output from piezoelectric transducer) to match the speed of conversion in a subsequent A/D converter. Similarly, when the final information concerning two measurands are related to their ratio or vector sum, appropriate processing has to be carried out prior to their display or recording.

12.9.1. Phase Sensitive Detection

In carrier wave measurements employing LVDT, information is contained in amplitude as well as in phase of the ac output. In such cases, conventional AC to DC conversion through rectification is not useful and a phase sensitive detection is required instead. In its simplest form, a phase sensitive detector consists of a multiplier which receives as input, the amplified signal from transducer and a reference sinusoidal source with fixed amplitude and with a phase identical to that of the ac excitation used. The output from the multiplier is filtered and fed to an indicator. The multiplier can be replaced by a FET used as a switch. The n channel FET Q1 is driven by a square wave signal synchronous with the excitation signal. The FET output is filtered and amplified further in another amplifier to give the dc proportional to the amplitude of the ac input signal. The disadvantage of using a switch (FET) instead of a multiplier is that the phase sensitive detector becomes sensitive to the fundamental as well as harmonics of the reference signal (3*f*, 5*f*, etc.). If this is not desired, switch can still be used provided the first

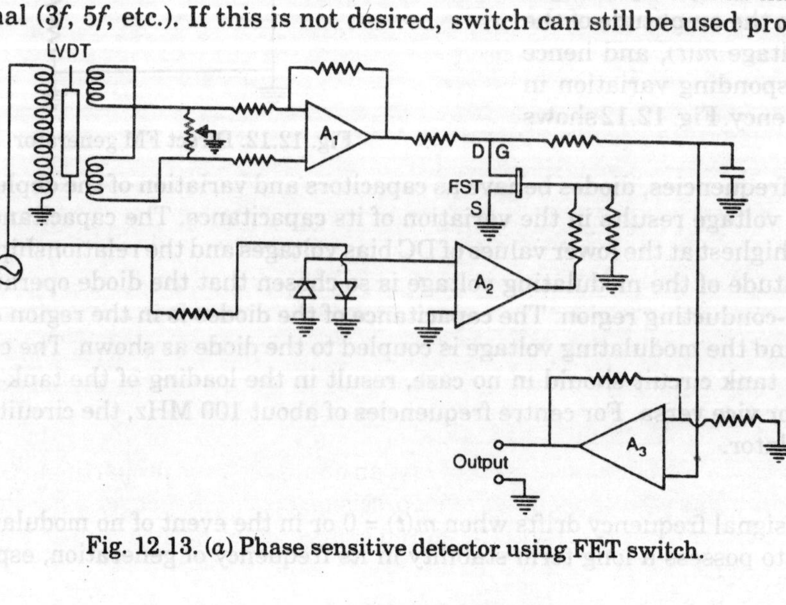

Fig. 12.13. (*a*) Phase sensitive detector using FET switch.

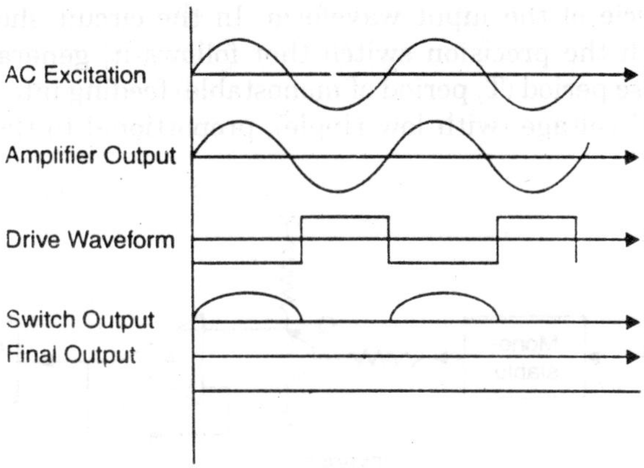

Fig. 12.13. (*b*) Output waveforms from phase sensitive detector.

amplifier is followed by a tuned filter, tuned exactly to the excitation frequency. The final filter can be chosen to be of a higher order, if required, to enhance the carrier frequency rejection at the output.

The basic circuit and the resulting waveforms have been shown in Fig. 12.13.

12.9.2. Comparators

The simplest comparator is an operational amplifier with high gain, operated in an open loop configuration. Due to high gain, a few hundred microvolts change in input can change the output level from negative to positive saturation. The maximum range of input over which comparison can be made, is limited by the allowable common mode voltage range. Additional input protection can be provided for inputs exceeding the differential input voltage limitations (Fig. 12.14). However, the saturation voltage is not immediately reached at zero crossing as the transition is limited by the slow rate which for the standard frequency compensated OA is of the order of 1 Volt/ms and for a supply of + 15V (assuming no offset error), the time required is about 15 × 2 × 1 = 30 μs.

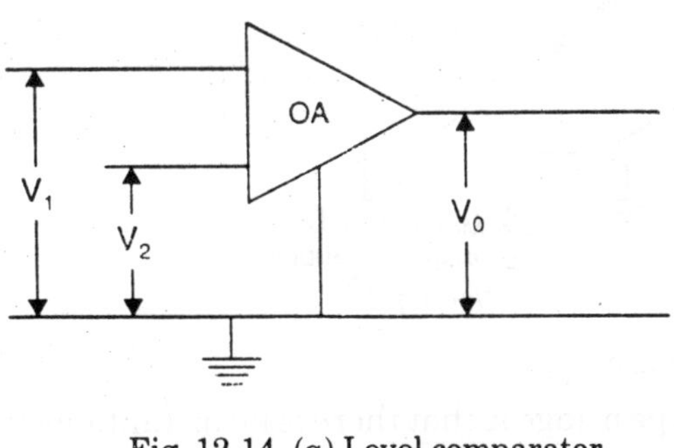

Fig. 12.14. (*a*) Level comparator.

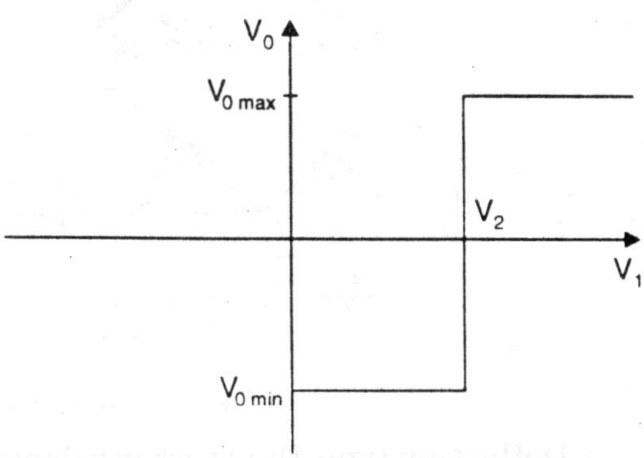

Fig. 12.14. (*b*) Level comparator characteristics.

12.9.3. Frequency to Voltage Converters

A frequency to voltage converter provides an analog voltage proportional to the frequency of the input signal. The circuit used is often a precision charge dispenser wherein a capacitor is charged to a predetermined level and the stored charge is discharged into an integrator or low pass circuit for every cycle of the input waveform. In the circuit shown in Fig. 12.15, the monostable together with the precision switch that follows it, generates a pulse of precise amplitude (V_R) and precise period (T, period of monostable) feeding into an averaging network. The final output is a DC voltage (with low ripple), proportional to the average of the input frequency.

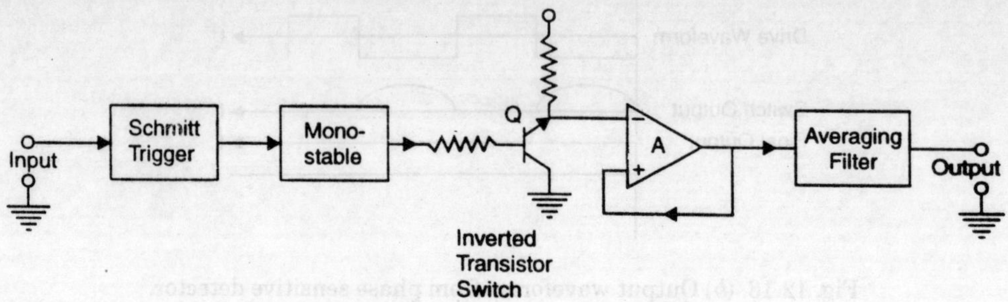

Fig. 12.15. Frequency to voltage converter.

12.10. BRIDGES AS INPUT CIRCUITS

AC or DC bridges are very often used for measuring the output of transducers. While resistive bridges are DC Wheatstone type, inductive or capacitive type bridges are used with AC.

12.10.1. DC or Wheatstone Bridges

(*a*) **Null of Self balancing type.** As shown in Fig. 12.16, variation in resistance R1 (transducer output) produces voltage output which is amplified and applied as an error signal to the field winding of a dc motor. Motor stops and slide wire on scale come to rest. Readout scale is calibrated in resistance change.

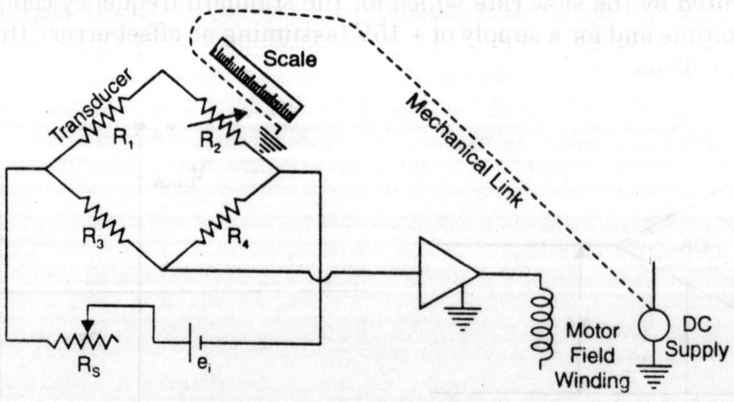

Fig. 12.16. Self balancing bridge.

(*b*) **Deflection type.** Drawback in balancing type bridge is that there is no instantaneous balance even if done automatically. Hence it is unsuitable for dynamic measurements, where changes are rapid. A null galvanometer is used for static inputs while CRO or recorder for dynamic inputs. In the latter case or when connected to VTVM or DVM, it is called voltage sensitive and when connected to indicating instrument, the bridge is called current sensitive. (Since these are low impedance devices)

Thus, in deflection type bridges (initially balanced), it is the out-of-balance voltage produced as a result of change is resistance produced by the tra.\sducer in one arm, which is measured and the calibration is in terms of the quantity under measurement.

There are some AC bridges for a push-pull arrangement of transducers. These accept input and work on differential principles. Their advantage is that these produce a more linear relationship with the bridge output and the measurand, as compared to a single element transducer bridge.

When two transducer elements are connected in adjacent arms of the bridge, any equal changes, *e.g.* those due to temperature, act against each other and get cancelled out. There is an inductive coupling between these adjacent arms of the bridge.

For transducers using high Q coils (*i.e.* $\omega L >> R$), change in the resistance of the transducer coils is insignificant in comparison with the change in inductance.

Therefore, output voltage $e_0 = \frac{1}{2} \cdot \frac{\Delta L}{L} e_i$

This is shown in Fig. 12.17.

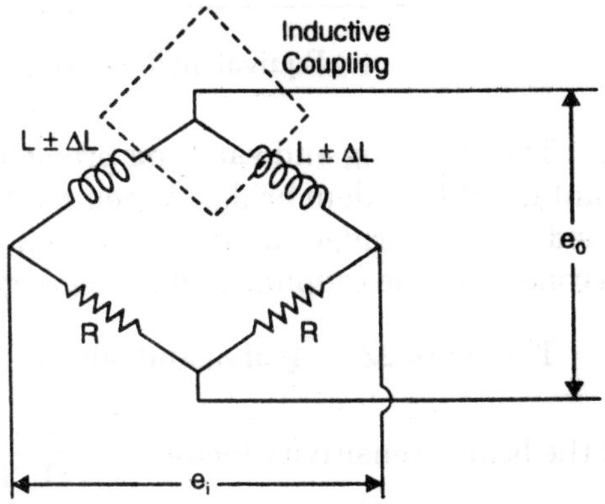

Fig. 12.17. AC bridge for push-pull transducer.

12.10.2. Inductive Transducer Blumlein Bridge

This type of bridge uses two closely coupled inductive ratio arms and two other arms consist of one part each of a push-pull transducer of a high Q. The bridge is shown in Fig. 12.18 (*a*) while its equivalent T network is shown in Fig. 12.18 (*b*).

From this : $Z_{AD} = Z_S + Z_P = j\omega M$

The coupling factor between the closely coupled coils is

$K = \pm M/LC$ and for perfect coupling, $K = \pm 1$.

When the bridge is balanced, currents in the two coils are equal and are in the same direction and for a particular winding sense.

Hence, $K = \pm 1$

With the balanced bridge, the impedance across the ratio arms will be

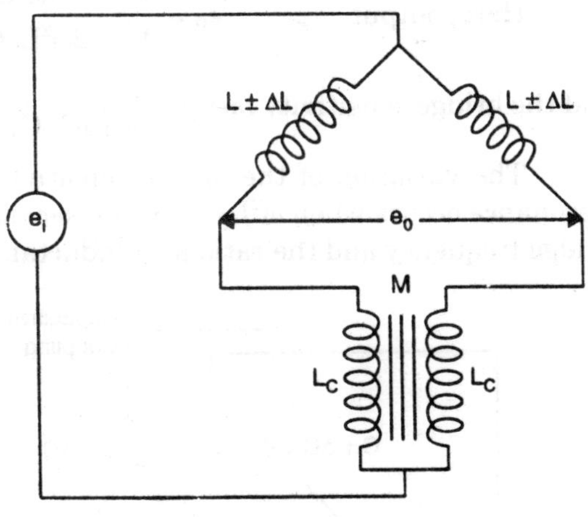

Fig. 12.18. (*a*) Blumlein bdidge for inductive transducers.

$$2Z_s = 2j\omega(LC - M),\ K = M/\sqrt{L_1}\sqrt{L_2}$$

when $K = 1$, $2Z_s = 0$

Thus there is no voltage drop across the ratio arms, and any stray or cable capacitance which is in parallel with the ratio arms, becomes ineffective.

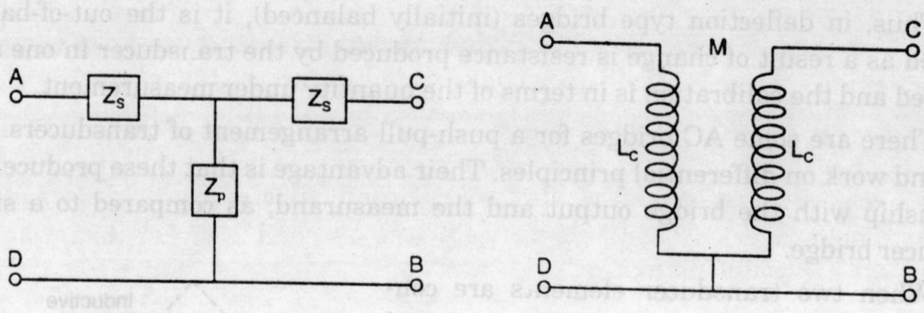

(*b*) Equivalent T-network. (*c*)

Fig. 12.18

The shielding and earthing problems are thus greatly simplified, which is the greatest advantage of Blumlein bridge. When the bridge is not balanced, the change in currents on the two sides of the bridge can be considered as the circulating current around the bridge and for this condition, the coupling factor k changes sign and becomes K = – 1.

Therefore $2Z_s = 4j\omega L_C$, and output $e_0 = \dfrac{\Delta L}{L} \times \dfrac{4L_C/L}{(1 + 2LC/L)} \cdot e_i$

and the bridge sensitivity factor $= \dfrac{e_0}{e_i\,(\Delta L/L)} = \dfrac{4L_C/L}{(1 + 2LC/L)}$

ΔL being too small $(\Delta L/L)^2 << 1$.

12.10.3. Capacitive Transducer Blumlein Bridge

This bridge is shown in Fig. 12.19 (*a*).

Here, output $$e_0 = \frac{\Delta C}{C} \cdot \frac{4\omega^2 L_C C}{2\omega^2 L_C C - 1} \cdot e_i$$

and the bridge sensitivity factor $= \dfrac{e_0}{e_i\,(\Delta C/C)} = \dfrac{4\omega^2 L_C C}{2\omega^2 L_C C - 1}$

The variation of the bridge sensitivity factor K with $\omega^2 L_C$ is shown in Fig. 12.19 (*b*). Resonance occurs when $\omega^2 L_C >> 1$, the sensitivity factor is 2 and is independent of variation of bridge frequency and the ratio arm inductance LC.

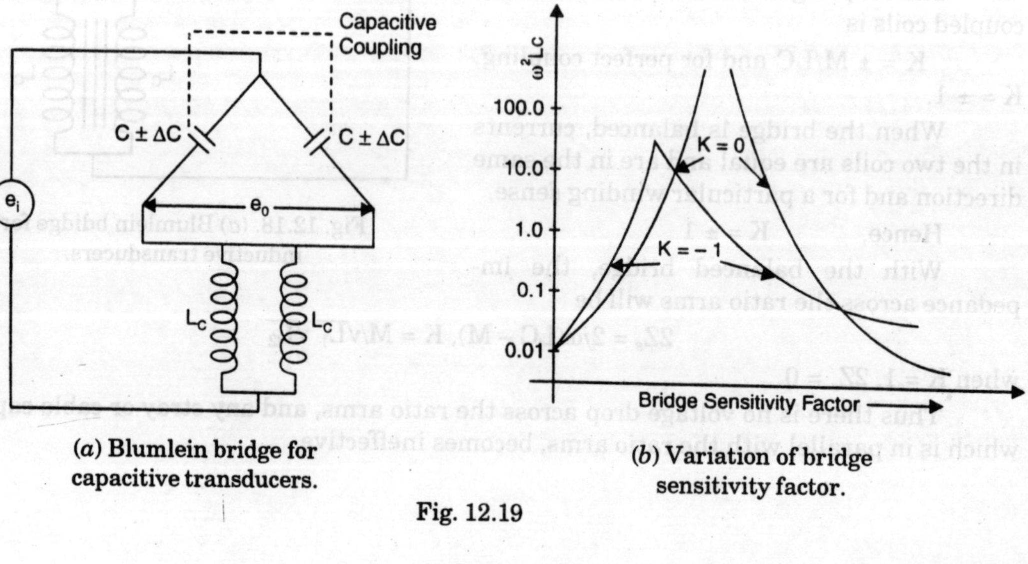

(*a*) Blumlein bridge for capacitive transducers. (*b*) Variation of bridge sensitivity factor.

Fig. 12.19

The uncoupled bridge (K = 0) has no horizontal characteristics and there is no region where sensitivity is independent of frequency or inductance, in that case.

Thus, advantages of a capacitive transducer bridge with tightly coupled arms are :

- No noise and peak up problems.
- Insensitivity to variations in frequency and inductance provided $\omega^2 L_C >> 1$.
- The bridge sensitivity is practically constant and is higher than that of the bridge with uncoupled arms.

BIBLIOGRAPHY

1. D.V.S. Murthy, *Transducers and Instrumentation Systems*, Prentice Hall of India Ltd. 1995.
2. D. Patranabis, *Sensors and Transducers*, Wheeler and Co. India, 1997.
3. T.D.S. Hamliton, *Handbook of Linear Integrated Electronics for Research*, Mcgraw Hill U.K. 1977.
4. J.G. Greene, *Applications of Operational Amplifiers, Third Generation Techniques*, Burr Brown Research Corporation, Mcgraw Hill, New York 1973.
5. ER Hnatek, *Applications of Linear Integrated Circuits*, John Wiley and Sons, New York, 1975.
6. *Application Manual for Operational Amplifiers*, Philbrick/Nexus Research, Dedham, Massachusetts 1965.
7. C.S. Rangan, G.R. Sarma and V.S.V. Mani, *Instrumentation : Devices and Systems*, Tata Mcgraw Hill, 1983.
8. U. Tietze and Ch. Schenk, *Advanced Electronic Circuits*, Springer Verlag Berlin 1978.
9. D. Patranabis, *Telemetry Principles*, Tata Mcgraw Hill 1999.
10. J. Das, S.K. Mullick and P.K. Chatterjee, *Principles of Digital Communication*, New Age International (P) Ltd. 1986.
11. T.R. Padmanabhan, *Industrial Instrumentation : Principles and Design*, Springer-Verlag 2000.

Index